Neues verkehrswissenschaftliches Journal

Ausgabe 29

Optimization of Schedules with Heterogeneous Train Structure in Planning of Railway Lines

Von der Fakultät Bau- und Umweltingenieurwissenschaften

der Universität Stuttgart zur Erlangung der Würde eines

Doktor-Ingenieurs (Dr.-Ing.) genehmigte Abhandlung

Vorgelegt von

Hyoung June Kim

aus Seoul, Südkorea

Hauptberichter: Prof. Dr.-Ing. Ullrich Martin

Mitberichter: PD Dr.-Ing. Yong Cui
Prof. Dr. rer. pol. habil. Georg Herzwurm

Tag der mündlichen Prüfung: 23. Oktober 2019

Institut für Eisenbahn- und Verkehrswesen der Universität Stuttgart

2019

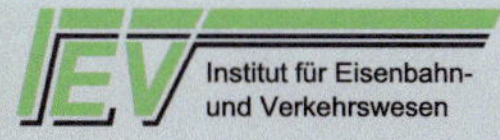

Herstellung und Verlag: BoD – Books on Demand, Norderstedt

Printed in Germany

ISBN 978-3-7494-8506-2

Preface

When designing railway infrastructure, it is frequently necessary to derive an initial, rough operating program based on traffic needs that acts as the basis for infrastructure dimensioning and takes into account defined, operational quality requirements. This process is described as capacity research and in recent decades, a variety of capacity research methods and procedures has been developed that are also successful in practical implementation. In particular, analytical methods used for investigating areas of homogeneous infrastructure, and simulation methods used for complex network structures have proved their worth.

In Germany and Europe, it is now rare to build new, longer railway lines to be used for mixed traffic with both slower and faster trains. However, in other parts of the world such as Asia, it is still an important component in both national and transnational transport system design. In such situations, the application of analytical methods is very limited because the results are extremely aggregated. When using simulation methods, numerous, detailed assumptions have to be made during the preliminary planning phase. A considerable amount of work is then needed to vary and optimize these assumptions within the framework of the capacity research.

In this dissertation, Mr. Kim develops a hands-on approach to determining optimized timetables based on a rough operating program for longer railway lines with mixed traffic. His approach, which requires a manageable amount of work, yields sufficiently detailed results that can also be used to draw conclusions about the proper arrangement of sidings.

The main research result of this work is a new, genetic procedure based on heuristic methods for the optimized design of an operating program on longer, mixed traffic railway lines. The results developed in this dissertation for optimizing timetables, especially for longer railway lines, based on a rough operating program and the derivable requirement-based arrangement of sidings are a considerable scientific contribution to the efficient planning of railway operations. The results not only close a gap between established analytical and simulation methods in railway operational science, but also have a direct relevance for the real design of railway systems.

Stuttgart, October 2019
Ullrich Martin

Dedication

I dedicate this thesis to my lovely wife HyeYeon Shin, who stood beside me with prayers and encouragement to achieve the vision that God showed to me, and to my three children, who sacrificed their needs to help their father complete this doctoral study, and to my parents who always pray for me to accomplish anything I set my mind to.

Acknowledgements

First of all, I give honor and glory to God, who steered me to accomplish this doctoral study.

I would like to express my sincere appreciation to Professor Ullrich Martin, head of the Institute of Railway and Transportation Engineering at the University of Stuttgart and Director of the Institute of Transportation Research at the University of Stuttgart. He gave me the opportunity to earn my Ph.D. degree in Germany and helped me from the beginning of the studies until I finished my thesis by providing me with advice and encouragement. I also thank PD Dr.-Ing. Yong Cui, my supervisor who advised me academically in my writing.

A tremendous thank you to my wife. This study would not have been completed without the prayers and loving support of my dear wife HyeYeon Shin. Finally, I thank my three children HaSeon, IHan and JiHan, who abandoned their needs to help me complete my Ph.D. degree.

Eidesstattliche Erklärung

Hiermit erkläre ich, dass ich diese Arbeit selbständig verfasst und keine anderen als die von mir angegebenen Quellen und Hilfsmittel verwendet habe.

Stuttgart, den 23.10.2019   Hyoung June Kim

Table of Contents

List of Figures

List of Tables

Kurzfassung

Aktuell wächst die Möglichkeit einer innerkoreanischen Eisenbahnverbindung, weil das Thema der Eisenbahnverbindung nach China über Sinuiju in Nordkorea bei den südkoreanisch-chinesischen Gipfelgesprächen diskutiert wird. Es wird erwartet, dass nicht nur Hochgeschwindigkeitszüge, sondern auch Güterzüge über Russland und China nach Europa gelangen können. Um diese Erwartung zu erfüllen, ist es notwendig, die alternden Bahnanlagen in Nordkorea zu verbessern. Bei der Planung vor dem Bau der Schienenwege muss jedoch, ausgehend von der verkehrspolitischen Aufgabenstellung, das Betriebsprogramm als Grundlage eines künftigen Fahrplans berücksichtigt werden. Wenn verschiedene Arten von Zügen auf einer Linie betrieben werden, ist eine Räumung der Durchgehenden Hauptgleise von Niedriggeschwindigkeitszügen für das Überholen durch Hochgeschwindigkeitszüge in Abhängigkeit von dem Zeitintervall, der Geschwindigkeit, der Zuglänge und der Entfernung zwischen den Stationen nicht vermeidbar. Die Analyse des auf dem Betriebsprogramm beruhenden Zugfahrplans in der Planungsphase vor dem Bau oder bei der Verbesserung einer Eisenbahninfrastruktur wird jedoch gegenwärtig oftmals noch auf der Grundlage der intuitiven Erfahrungen und Kenntnisse des Planers durchgeführt, und die Optimierung des Zugfahrplans unter verschiedenen Bedingungen wird nur ansatzweise berücksichtigt.

In dieser Studie wurde ein Modell zur Bestimmung eines optimalen Zugfahrplans in der Planungsphase, beim Einsatz dreier Arten von Zügen mit unterschiedlichen Geschwindigkeiten, Betriebsfrequenzen und Halten auf einer Eisenbahnlinie unter Berücksichtigung des Zugfahrplans entwickelt. Die Netzwerkmodelle können je nach Methode in Raum-Zeit-Netzwerk, Raum-Netzwerk und Standort-Zeit- kategorisiert werden. In dieser Arbeit wird ein auf dem Standort-Zeit-Netzwerkansatz beruhendes mathematisches Modell vorgeschlagen, das eine Reihe von Vorteilen hinsichtlich der Verwendbarkeit verschiedener Zielfunktionen und Nebenbedingungen, der Modellskalierbarkeit, einer einfachen Implementierung sowie der Interpretation der Ergebnisse bietet. Da Zugfahrpläne als NP-schweres Problem bekannt sind, dessen Lösung aufgrund einer großen Anzahl von Entscheidungsvariablen sehr oft nicht innerhalb eines angemessenen Zeitraums erreicht werden kann, werden in vielen Studien zur Entwicklung eines Zugfahrplans heuristische Methoden angewendet.

Um die Zeitverzögerung bei der Räumung Durchgehender Hauptgleise von Niedergeschwindigkeitszügen für die Überholung durch Hochgeschwindigkeitszüge zu minimieren und um zu bestimmen, welche Bahnhöfe Überholgleise benötigen, wurden die sich daraus für den Zugfahrplan ergebenden Restriktionen als Nebenbedingungen angenommen, und Methoden, die möglichst viele dieser Nebenbedingungen abbilden können, wurden überprüft.

Im Ergebnis stellte sich heraus, dass ein heuristischer Algorithmus für die hier beabsichtigte Optimierung am besten geeignet ist. Folglich wurden verschiedene heuristische Algorithmen evaluiert und ein genetischer Ansatz wurde für diese Studie aufgrund vielfältiger Vorteile, z.B. der Skalierbarkeit des Modells, der einfachen Integration und Relaxation von Nebenbedingungen sowie der Möglichkeit, verschiedene Nebenbedingungen zu berücksichtigen, ausgewählt. Zur Verifizierung der auf dem entwickelten mathematischen Modell beruhenden Optimierung von Zugfahrplänen wurde der verwendete genetischer Algorithmus auf einer Linie in Südkorea, die sich in der Planungsphase befindet, angewendet.

Der in dieser Arbeit verwendete genetische Algorithmus analysiert zufällig generierte Zugfahrpläne (Chromosomen) über 300 Generationen, und der Zugfahrplan (Chromosom) mit der besten Eignung (kürzeste Verzögerungszeit) wird als optimaler Zugfahrplan (optimale Lösung) erkannt. Im Modell wird eine Analyse hinsichtlich der besten Eignung (beste Fitness) entsprechend eines gewählten Zielwertes, der Anzahl der Räumungen von Durchgehenden Hauptgleisen, die an den jeweiligen Bahnhöfen auftreten, und der Anzahl der Zugfahrpläne mit identischen Überholungsbahnhöfen durchgeführt.

Durch die Optimierung des Zugfahrplans auf einer geplanten Eisenbahnlinie und Festlegung der Überholungsbahnhöfe nach der in dieser Studie vorgeschlagenen Methode ist es möglich, eine objektive Grundlage für die Bahnplanung zu nutzen, statt sich allein auf die subjektive Erfahrung der Planer zu verlassen. Darüber hinaus kann ein so optimierter Zugfahrplan auf der Grundlage der in dieser Studie vorgeschlagenen Methode analysiert wird, kann dies dazu beitragen, die wirtschaftliche und politische Durchführbarkeit des Baus neuer Eisenbahnstrecken zu erreichen.

Abstract

Currently, the possibility of the inter-Korean railway connection is increasing as the topic of the railway connection leading to China through Sinuiju in North Korea is discussed at the Republic of Korea-China summit. It is expected that not only high-speed trains but also freight trains could reach beyond Russia and China, to Europe. In order to realize this expectation, it is essential to improve the aging railway facilities in North Korea. One of the most important things to consider when planning facilities before constructing a railway is the train operating program. Furthermore, when different types of trains operate in one line, the following high-speed train to overtake the preceding train take places depending on the train time interval, speed, length of trains, and distance between stations. However, the analysis of the operating program based train schedule in the railway planning stage for the construction or improvement of railway infrastructure is carried out mainly on the basis of the intuitive experiences of the planner, and the optimization of the train schedule under various conditions is not considered properly.

Considering the importance of the train schedule in the railway planning stage, this study developed a model for analyzing the Optimal Train Schedule (OTS) when three types of trains with different train speeds, operation frequencies, and stopping stations operate in one railway line. The existing railway models can be categorized into space-time, space and location-time models depending on the expression method. In this thesis, a mathematical model based on the location-time model, is proposed, which has many advantages in terms of acceptability of various objective and constraint functions, model scalability and ease of implementation and ease of communication such as interpretation of results. Furthermore, because railway train schedules are known as an NP-hard problem whose solution cannot be derived within a reasonable period of time owing to a large number of decision-making variables, most studies on the development of a train schedule problem model use the heuristic methodology.

In order to minimize the scheduled waiting time for overtaking low-speed trains due to the following high-speed trains and determine the stations that require a Subsidiary Main Track (SMT), the conditions required for planning a train schedule were expressed as constraints, and the methods that can accommodate as many of these constraints as possible were reviewed. As a result, it was determined that the heuristic method is the most appropriate for the optimization research in this thesis. Consequently, various optimization algorithms were compared, and a genetic algorithm was selected for this study because of its various advantages such as scalability of the model, ease of addition and relaxation of constraints, and possibility of accommodating various constraints. To implement the optimization of the train schedule based on a mathematical model, a genetic algorithm was developed and applied to a line during the planning stage in Republic of Korea.

The genetic algorithm used in this study analyzed random train schedules (chromosomes) over 300 generations, and the train schedule (chromosome) with the best fitness (lowest scheduled waiting time) becomes the OTS (optimal solution). An analysis was performed on the best fitness according to target fitness, the number of overtaking occurred at each station, and the number of train schedules with the same overtaking stations.

If the OTS is established by applying the method proposed in this study to the planned railway line, and the stations where low-speed trains can be overtaken by high-speed trains when trains of different types operate in one railway line are determined in accordance with the established the OTS, then it will be possible to provide an objective basis for railway planning, rather than relying on the subjective experience of the planners. Furthermore, if the OTS is analyzed based on the method proposed in this study when a new railway line is constructed in a developing country, then it will help achieve economic and policy feasibility.

1 Introduction

Road traffic congestion in many towns and cities has increase owing to urbanization, population growth, existing narrow roads, limited investments in roads, and the increasing number of cars on the roads. This has resulted in an increase in research on sustainable public transport systems as it has become increasingly important. The efficiency of public transportation systems is continuously researched with a primary focus on the operation level and maintenance level.(Oh 2012).

Globally, there is a paradigm shift in transport systems as the investment in railway systems has increased in comparison to road transport systems. A Safe, efficient, economical, and comfortable railway transport realized through high-speed trains has increasingly become popular (Wang 2018).

Railway transportation systems have shown that they are potentially green mode of public transport (Cacchiani et al. 2014). The aforementioned benefits of the railway transport have led to the construction of new rail tracks and the repair of existing rail tracks in many countries, however, it requires a long-term planning. Railway planning proceeds decision-making process in a way step by step. It is characterized by a strategic, tactical, and operation level as shown in Figure 1. A detailed process for each level also corresponds with the classification system of railway planning (Zimmermann and Lindner 2003, Liebchen and Möhring 2007, Lusby et al. 2011).

Railway planning at the strategic level is characterized by a long-term focus on the infrastructure and can be subdivided into demand analysis, network planning, and line planning. The demand for passengers and cargos constitutes the basic data for railway network planning, line planning, and timetable generation.

Economic growth widens the range of human activities and increases trip frequency, therefore, traffic demand should be surveyed for railway planning (Do 2013). The demand for passenger and freight are made at a long-term strategic level and are analyzed using origin-destination surveys. The surveys are designed to gather data on the number and type of trips in an area, are the data is fundamental for railway network planning, line planning and timetable generation (Schroeder et al. 2010).

Railway planning at a long-term strategic level focuses on the construction of railway networks and the modification and supplementation of the existing railway networks. It primarily focuses on the potential of securing the capacity for a future railway network at the time of planning.

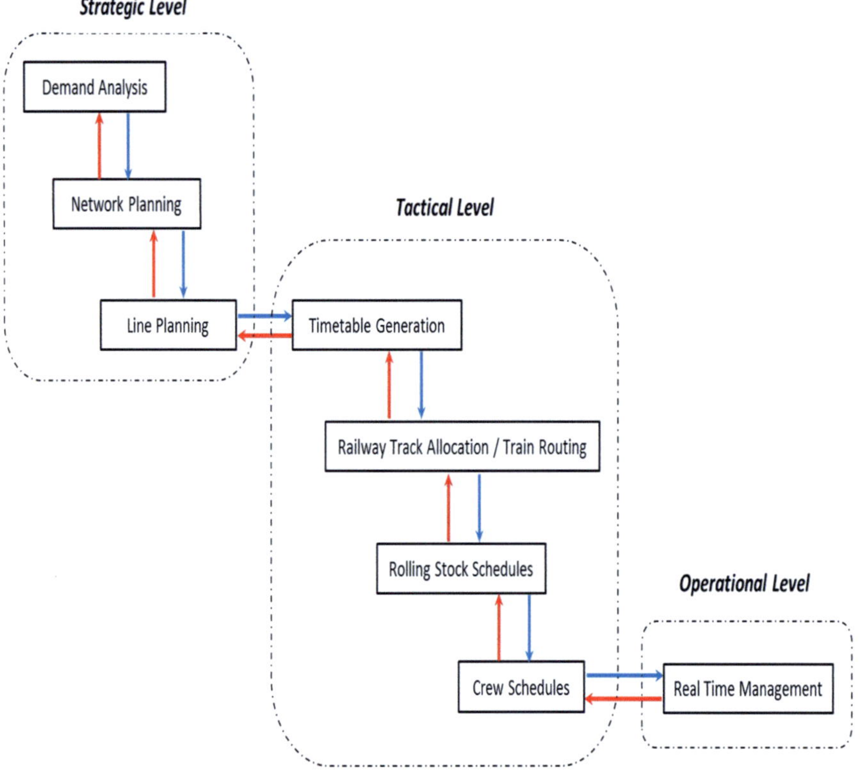

Figure 1: **Railway Planning Process (source: Zimmermann and Lindner 2003; Liebchen and Möhring 2007; Lusby et al. 2011)**

Railway line planning involves dividing railway services into different classes under the secured capacity and developing plans (including the start and end points, and frequency of services) for the lines to allow for provision of sufficient railway services for each class.

Railway planning at a tactical level focuses on the distribution of the resources of the constructed railway infrastructure and can be subdivided into timetable generation, track allocation and train routing, rolling stock scheduling, and crew scheduling. The timetable generation determines the stations where a train is to start from and make its final stop i.e., start and end stations. Each train must have start and end station times including the arrival times at each station where it stops on each line it uses. This must not exceed the line capacity of the section between specific stations. In general, passenger train operators repeatedly use cyclic (or periodic) timetables that have a set pattern and a set headway to prevent services from exceeding the line capacity (Lusby et al. 2011). Hence, the provision of transfer services for the railway lines of other operation companies as well as their own, train operators match the departure and arrival times between linked trains at specific stations or ensure good dwell times for the trains at the transfer stations. Potential challenges due to line capacity must be considered during the train operation planning level, and the output of a train service plan is the train schedule. If for example a train would like to overtake the preceding train on one railway track, it may only be possible if there is a relief track or secondary main tracks inside a station. The main track is the principal portion of a rail transit line used by trains when arriving, departing and passing through a station. Based on its intended purpose, the main track is divided into an up track, a down track, an arrival track, a departure track, a passenger track, a freight track, and a through track (for trains passing through), and other principal main tracks are regarded as Subsidiary Main Tracks (SMTs). The challenges with track allocation are overcome by identifying conflict-free train routes of maximum value for a given railway network. Railway track allocation is considered because the train routing on a railway network entails allocating the track capacity of the network in a conflict-free manner (Lusby et al. 2011).

Railway planning is considered at a tactical level and includes the rolling stock schedule, crew management including schedules and train schedules. Rolling stock allocation refers to the allocation of rolling stock units to meet the train timetables that may have been set based on the rolling stock base (Lusby et al. 2018). Crew management is the distribution of personnel to guarantee that each train is equipped with the necessary operating staff (Zimmermann and Lindner 2003).

The existing literature includes studies on rolling stock circulation models that consider reality (Fioole et al. 2006), ; the train timetables of railway systems; location and overtaking of the locomotives and the operations parking yard owned by railway companies (Vaidyanathan et al. 2008). Crew allocation refers to the scheduling of a duty roster that shows the duties to be performed by each crew member and is comparable to airline crew scheduling (Lusby et al. 2018).

In the scope of this thesis, overtaking is defined as the passing through of a high speed train at station as it passes a preceding low-speed train as it stops on secondary main track in the station such that it does not obstruct its operation. After the high-speed train passes through the station, the low-speed train can depart. The dwell time of the low-speed train is increased to allow the high-speed train to pass through the station, hence, the overall waiting time (consecutive delay) is also increased. The dwell time for the transfer service and the secondary main tracks used by the overtaking train must be adjusted by the station tracks. The headway in a station is determined by the track arrangement status of the station.

Railway planning at an operational level focuses on measuring the management of various challenges in real time, which occur during the execution and operation level after planning the railway line usage. The primary challenge is the temporary modification of the timetable to timeously return to the original train service plan in response to real-time deviations such as delayed train service, urgent line repair and accidents. Hence, studies have been conducted on the readjustment of the train timetable, adjustment of the rolling stock allocation and adjustment of the crew schedule (D'Ariano et al. 2008, Rezanova and Ryan 2010).

Railway line planning at a strategic level is based on the expected passenger and freight demand. To overcome challenges with line planning, the throughput of passengers and freight can be increased whilst decreasing costs. However, researches on train schedules upon attendant timetable generation at the tactical level are not enough and is ongoing. The train scheduling planners for train operating companies propose train schedule based on their experience and expertise for planning train operation which is different from person to person. The proposed train schedules are related to building a new facility in the process of feasibility studies, basic planning, and implementation for extensive investment in railway projects.

This practice often leads to several limitations related to uncertainty and variability in train schedule planning. The train schedule is, after all, a matter of order of trains. It also includes placing secondary main tracks that allow waiting for another train to maximize railway capacity in case that different kinds of trains share the rails.

In the Republic of Korea, the mathematical method for rail capacity calculation is mainly based on a planners' intuitive experience and linear methods (Kim et. al 2017). The arrangement of trains on each line is set after assessing a number of probable arrangements through trial and error. These methods have limitations, as they do not thoroughly consider train arrangement optimization processes that minimizes the operation intervals of trains under various conditions. Railway planning is importance for the effective management at the phase of planning level (Bussieck et al. 1997, Lusby et al. 2011).

It is necessary to analyze capacity and scheduling robustness in railway planning in order to provide maximum transportation service with limited railway resources and improve railway management. The main objective of capacity research is to find out the interrelationship between the load (usually expressed as the amount of trains per unit of time) and the operation quality of the investigated system. It is an important method to evaluate the operating performance of railway operation on the infrastructure, checking whether the infrastructure is efficient utilized (Cao 2017).

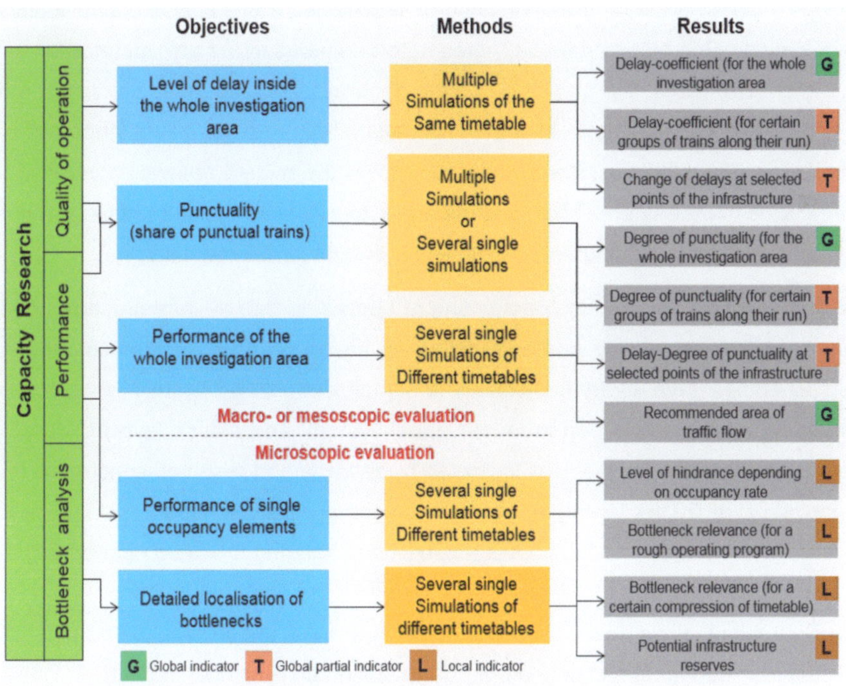

Figure 2: **Structure of Capacity Research (source: Cao 2017 and Martin et al. 2012)**

Figure 2 is represented the general structure of capacity research for railway systems. Capacity research can not only carry out macro- or mesoscopic evaluation of the railway operation in the whole railway network, but also evaluate the operation on a certain occupancy element microscopic. The level of delay and punctuality inside the whole investigated area are two major items to improve quality of operation. Multiple simulations of same timetables with perturbation are used to check the robustness of timetables regarding delay propagation. The performance of the whole investigated area is another important objective to evaluate the quality of operation and the capacity of an investigated area. For microscopic evaluation of homogeneity, the performance of single occupancy elements is evaluated through several single simulations of different timetables in order to identify the bottlenecks in the whole network. (Cao 2017).

Railway capacity is determined by factors which include train heterogeneity, train operation patterns, infrastructure layout, the speed of trains and train scheduling. It is an important factor as it determines the ability of stations to allow a low-speed train to stop for high-speed trains through analyzing train schedules. Furthermore, the location of stations that can allow for the overtaking of slow-speed trains is also a significant element in evaluating railway capacity when heterogeneous trains are operated on a railway track. Therefore, line planning and the generation of timetables are key in determining railway capacity as well as train allocation, train routing, rolling stock and crew scheduling. The aforementioned steps determine the overall railway networking and its real-time management. Thus, it is important to establish optimum railway planning that considers line planning and timetable generation simultaneously.

Railway planners involved in constructing or upgrading railway infrastructure conduct train operation scheduling primarily manually, based on their experience. However, train schedules have a very fastidious and complex structure because of various constraints (such as vehicle performance, signal system, and station layout). Moreover, the evaluation of the time table is impacted by capacity and resource constraints (Hassannayebi et al. 2016). Owing to these challenges, the train schedule changes require time and once a train service plan is established changing it can be challenging. Furthermore, if the constraints are changed, such as the changes in the operation time (i.e., hour/minute) and headway due to a facility change, the timeous preparation of an efficient train schedule can only be performed using an empirical method.

The inter-Korean summit, has risen expectations with regards to the possibility of an inter-Korean railway system that will connect South Korea and China through North Korea. It is anticipated that the inter-Korean railway system will be discussed during the Republic of Korea–China summit, and if approved the North Korean railway facilities will be upgraded owing to their aging railway infrastructure. Once upgraded it is expected that not only high-speed trains but also freight trains would run from Korea through Russia and China to Europe.

This thesis analyzes optimal train schedules (OTS) in double track lines based on the frequency of trains with different characteristics that was calculated based on traffic demand. Systematic ways should be suggested to decide combinations at an acceptable level about the optimized train schedule and placing of track arrangement for operation of heterogeneous train in railway planning. In this study, metaheuristics were used to perform objective functions in mathematical models based on location-time models to determine the SMT locations for overtaking the railway planning process between the strategic and tactical level. The results of this study will contribute to the railway plan for the anticipated connection between the South Korean railway line and the North Korean line.

The scope of the present study is described in four parts. Firstly, the study focuses on inter-regional passenger railway system and not on urban railway systems that operate over short distances. Inter-regional train operate over long distances between stations, have long headways, and use various types of trains with different characteristics. Thus, the planning of the train schedule is of great importance. The benefits of improved efficiency at an operation level will be greater for the inter-regional train in comparison to urban trains.

Secondly, in inter-regional railway systems different types of trains can operate on the same lines and the stops of each train are predetermined. This is because the types of trains operated on each line and the use of selected stations are generally determined by policy. Hence, in this study a model for analyzing the stops of low-speed trains when high-speed trains overtake them to achieve the OTS and efficient train operation in a case where the train operation frequency and the stop stations of different types of trains are already determined.

Thirdly, it is assumed that the demand between railway stations does not change. In principle, the inter-zonal traffic demand of the railway must be changed through the modal split with other transportation modes depending on the traffic time and cost of the railway. However, inter-regional railways have fewer changes in demand owing to low sensitivity to the modal spilt. This study analyzes the OTS by setting the train operation frequency as the input based on traffic demand.

Fourthly, the beginning and ending pattern of the line is not fixed but calculated internally by the model, and it is assumed that there are no transfers. To consider transfers, the synchronization time for transfers of passengers must be considered, but this presupposes the preparation of the train schedule. This study focuses on the aspects prior to the preparation of the train schedule, hence, transfers were not considered.

The safe and effective train operation depends on constraint conditions such as train distribution at stations, the number of platforms, the number of trains entering (either to stop or to pass) a station, and safe distance between trains. This study aims to overcome the limits that the past research had as follows:

- Analysis for optimum train schedule for heterogeneous trains which have different speeds, stopping stations and traffic demands for respective trains in a railway planning process.
- Three train schedules are presented after analyzing the scheduled waiting time by overtaking at each station.
- Finally, Suggests stations which need SMTs for overtaking based on optimum train schedule analyzed by the genetic algorithm.

The scope and purpose of this study are set up as described above, and the meaning and relevance of each will be reviewed again in the main discussion.

The structure of this thesis is as follows. Chapter 1 gives the objective and the scope in this thesis. Chapter 2 examines the basic principles in train schedule, which constitute the theoretical background of this study. Chapter 3 presents various optimization methods are introduced, and the method applied in this study and the research flowchart are presented. And the methodology for implementing the optimization and it mentions how to make genetic algorithm processes. Chapter 4 introduces mathematical models for train schedule in literature review. The general matters that should be considered when analyzing a train schedule are described, and mentions the space-time network, space network, and location–time models, which depend on the railway network expression method to find the optimal solution. Furthermore, the research focus and novelty of this study are presented after reviewing related works. Chapter 5 presents a mathematical model for various variables, objective functions, and constraints for the train schedule analysis. Chapter 6 describes the methodology and procedure for developing the optimization algorithm. Furthermore, based on the mathematical model proposed in this study, the development of an optimization algorithm for a heterogeneous train schedule is realized. And it is described the Unified Modeling Language (UML). The results of implementing the method with a computer programming language, according to the research flowchart based on a genetic algorithm to meet the objective functions and constraints presented in this study, are presented. In Chapter 7, the developed optimization algorithm is applied to a railway line for evaluation, and analysis results are discussed. Chapter 8 presents the summary, conclusions and future research. The references are included in Chapter 9 and the detailed data in this study is presented in appendix.

2 Basic Principles of Train Scheduling

2.1 Meaning of Train Scheduling

Train scheduling means coordinating the train paths ordered by competing train operation companies. On railway networks that are operated on an open access basis, scheduling is not just a planning procedure but also the commercial interface between infrastructure operators and train operating companies. And the functions of scheduling performance are as follows (Pachl 2014).

- It coordinates the train paths in the planning process for optimum use of the infrastructure.
- It ensures the predictability of train traffic
- It produces timetable data for passenger information.
- It is essential for traffic control, locomotive and rolling stock usage and crew scheduling.

Railway planning is a process of determining a train schedule (cyclic timetable) that minimizes the cost of a train operating company while satisfying a series of physical constraints and increasing the satisfaction of passengers and shippers based on pre-established transportation plans. (Pachl 2014) presented that the scheduling process is carried out by manual scheduling and computer-based scheduling. Manual scheduling was the archetypal scheduling method for more than 150 years. Computer-based scheduling methods have been implemented after the appearance of high-performance computers in the 1990s. The blocking time model that represents train paths is used in some computer-based scheduling, which adopted fundamental concepts of manual scheduling. Therefore, it is still valuable for a description of train pathway modeling in manual scheduling.

The train schedule design is one of the most complex problem that a railway company must solve and a critical issue that affects various aspects of the railway company. Once a train schedule is designed and applied to the field, it becomes a job that must be performed by all employees of the railway company. Moreover, how the train schedule is designed determines the economic performance and quality of traffic service of a railway company. Along with the recent economic growth, the traffic demand for diversity of railway operation service is increasing. Many studies are under way to improve operation quality and solve the capacity limit[1]. In particular, (Cao 2017) presented homogeneity of operating programs based on infrastructure occupancy. In order to determine and evaluate homogeneous level of train traffic, the occupancy of infrastructure has been considered with speed, running time and headway. The blocking time, buffer time and running direction describe the occupancy of train path on track sections based on the blocking time model. Accordingly, the homogeneity of operating programs was evaluated through the homogeneity of blocking time (HBL), the homogeneity of buffer time (HBU) and the homogeneity of running direction (HRD). It was analyzed by RailSys[2] and PULEIV[3]. The result quantified the homogeneity of railway operation for track sections as well as an entire network. And the interrelationship between homogeneity of operating programs and operation quality were investigated. Finally, the HBU and HBL were studied regarding the effects of the low traffic flow and the high traffic flow in the homogeneity of operating programs.

A general train schedule that has been used for the longest time is the fixed train schedule. With the fixed train schedule, the starting time of each line is predetermined, and it is provided in a printed form that passengers can obtain easily. Passengers adjust their starting time to the available times listed in the schedule pamphlet. The railway company must abide by the train schedule announced to the passengers and the start trains at the predetermined times regardless of how many passengers have boarded the train. This type of fixed train schedule means that all activities of the railway company must be performed in accordance with the schedule. The fixed train schedule is presumed to last for a long time. However, the passenger flow and distribution may change over time. One problem with the fixed train schedule is that it cannot accommodate the passenger demand if the number of passengers increases sharply.

---

[1] Capacity Limit is a typical value that if the infrastructure occupation (% of time-window) is higher than or equal to the analyzed line section shall then be called congested infrastructure and no more additional train paths may be added to the timetable (UIC 2013).

[2] RailSys is a synchronous microscopic simulation program for railway system, which is developed by Rail Management Consultants, Germany.

[3] PULEIV is a software tool developed by the Institute IEV, University of Stuttgart, to investigate the performance behavior of railway networks.

A dynamic train schedule means organizing trains without fixing the train schedule. The primary purpose of using a dynamic train schedule is to adjust the starting times to satisfy the passenger demand to the maximum to provide high-quality services in the transportation market. The dynamic train schedule generally does not allow the reservations of seats a longer time period before a train starts.

This study is related to the analysis of fixed train schedules because it calculates the frequency of train operation for different types of trains on the basis of traffic demand and analyzes the train schedule by setting the departure time at a predetermined calculation time according to the frequency of each train.

The train operation frequency has a great effect on the modal split. The magnitude of this effect increases especially for a line in which various modes of transportation compete. The factors that travelers consider important when selecting railway are travel cost, travel time, operation frequency, and departure time. The share of railway transport can be increased by closely observing the travelers and its time distribution, and providing a sufficient number of trains at the right time (Ortúzar, Willumsen 2001). However, if the number of trains is insufficient or the departure times are inappropriate, a large part of the demand will shift to other modes of transport. One factor that influences the operation frequency is passenger flow. Passenger flow refers to the number of travelers per unit time who move from one city to another. Passenger flow is a function of time, because it varies over time. The variation in passenger flow can be observed over various periods (hourly, daily, weekly, and monthly). The daily and hourly variations of passenger flow are very important. In particular, the hourly variation plays a key role when determining the operation frequency and departure time (Ahn et al. 2017). The number of passengers per train must be considered when deciding on the train operation frequency. Furthermore, the train operation frequency is determined by the predicted traffic demand of the railway line divided by the train capacity to be operated on that line.

To determine a train schedule, the passenger demand should be considered, the economic performance of the railway company, and various operational constraints. The train schedule problem is essentially a combinatorial problem. In other words, every case should be examined individually and calculate how much each case satisfies the interests of railway companies, passenger demand, and operational constraints before choosing the most satisfying case. If the train operation frequency increases and the actually available departure times become more diverse, then the average time loss per passenger will decrease. Because the train operation frequency has a direct effect on the difference between the departure time of the train that a passenger initially selected and the departure time of the train where the passenger can guarantee a seat, it is the most important factor that influences the quality of the transportation service.

When determining the train operation frequency, the influence of train operation frequency on the quality of the transportation service must be considered. The measure of quality of operation is defined by waiting time that result from hindrances during the operational process (consecutive delays). The waiting times strongly depend on the number of trains in the network and train headways. The more trains occupy the infrastructure, the higher is the sum of waiting times. The more the sum of waiting time increases, the more the graph of the function coverages toward the (maximum theoretical) timetable capacity. The ratio of the amount of waiting time per train improves the comparability of different infrastructure or timetable options (Pachl 2014). Figure 3 represents the waiting time diagram. The waiting time function represents the average waiting time per train that occurs for a certain traffic flow. The waiting time is a measure of the quality of operation (Pachl 2002).

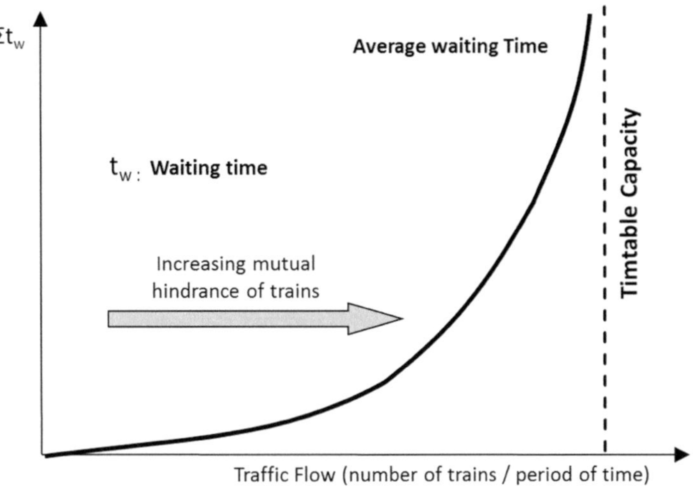

Figure 3:   **The Waiting Time Diagram and Capacity (source: Pachl 2014; Kim et.al. 2017)**

In the past, the expression of waiting time in railway networks was mainly used for the investigation of delays and delay propagation. However, a second category of waiting times can also be assessed in railway networks, the so-called scheduled waiting times (Wendler 2007). Scheduled waiting times arise during timetabling when different Train Operating Companies (TOCs) compete for the railway infrastructure available (Pachl 2014). In addition, scheduled waiting time is necessary when different kind of trains run on the same railway infrastructure because an artificial increase in the overall times of train schedules which is caused by conflicts of heterogeneous trains during scheduling process is an inevitable consequence. The scheduled waiting time generally leads to an increase of total time (longer stopping time and dwell time respectively, in some cases longer running time). (Pachl 2014).

This thesis calculates the operation frequencies of three types of trains (high speed, general, freight) on a train line for one day based on the analysis of traffic demand, and the train schedule optimized using a genetic algorithm is analyzed for minimizing weighted scheduled waiting time.

2.2 Basic Constraints for Train Scheduling

The basic constraints that must be considered in the train scheduling include starting station, terminating station, departure time, stopping stations, and dwell time in each station.

- Travel time: Sum of all times of a passenger' from origin to final destination of a trip including access times.
- Dwell time: The total elapsed time from the beginning of a scheduled train stop in a station until the time it resumes moving.
- Train meet in a single line: Trains meeting in a single line occurs when two trains travel in the opposite directions in a section connected by a single line. When two trains travel in the opposite directions, if one train first enters a section between two stations connected by a single line, the other train must wait until the first train passes, which causes scheduled waiting time.
- Overtaking: When two trains run in the same direction at different speeds and the low-speed train is in front; the low-speed train will be passed by a train with higher speed at a certain part of infrastructure.

- Minimum headway time: Minimum inter-arrival time between two train paths respecting the requested speed profile of both trains. The minimum headway time is determined in the first block section of that part of the infrastructure which is used in common by both trains. The first block section is not necessarily the relevant block section where the blocking times would touch without tolerance.
- Train meet in a station: Railway facilities consist of stopping stations and a line network connecting the stations.

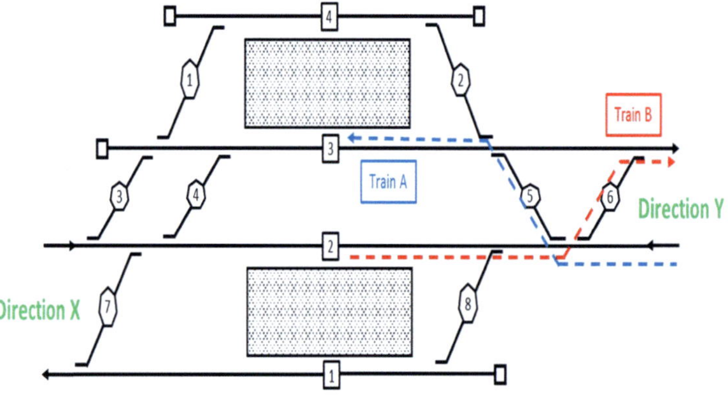

Figure 4:   **Example of a Stopping Station**

The stopping stations consist of multiple tracks and platforms. Figure 4 shows an example of a stopping station with 4 lines, 2 platforms, and 7 crossovers. In this figure, ☐ indicates the line number, ⭕ indicates a crossover, and the stopping stations are connected in double lines with neighboring stations. Train A, which comes from direction Y, passes crossover no. 5, stops at or passes line no. 3, passes crossovers no. 4 and no. 7, and travels to direction X. Train B, which comes from direction X, stops at or passes line no. 2, passes crossover no. 6, and travels to direction Y. For trains A and B to avoid a collision, they must satisfy the predetermined minimum time separation when they pass the same line. Allocation of the same line: If two trains close to each other in time are allocated to the same line in a station, the arrival times of these two trains must meet the minimum headway. In addition to these basic constraints, this thesis analyzes the Optimal Train Schedule (OTS) by adding constraints that occur when different types of trains run on one route using the methodology of genetic algorithms.

3 Optimization Methodology

3.1 Theoretical Background

The optimization problems can be broadly classified into problems with continuous and/or discrete variables. The problems with discrete variables are called combinatorial problems. In a continuation problem, the aim is typically to find a set of real numbers; in a combinatorial problem, the goal is a finite set (integers, permutations, graphs, etc.). Furthermore, there are mixed problems that have both continuous and discrete variables.

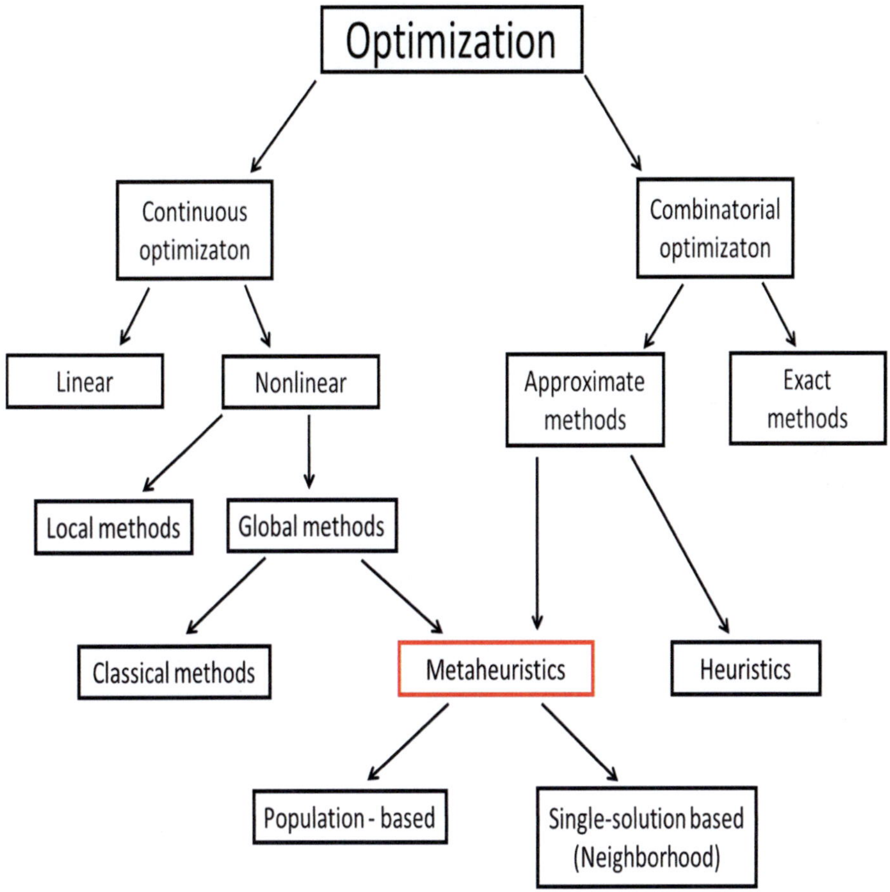

Figure 5:   **Classification of Optimization Methods (source: Kim 2017a)**

The optimization techniques for solving continuous problems can be largely divided into linear and nonlinear methods.

The local method is the oldest and simplest optimization technique. The local method arbitrarily sets one of the potential solutions as the initial solution and this initial solution becomes the current solution. This method finds a solution that improves the objective function in the neighborhood of the current solution. If such neighbor solution is found, set it as the current solution. This search process ends when there is no better solution in the neighborhood. When the local method is applied to a problem with multiple local optima, the search process is terminated when one local optimum is found. The disadvantage of the local method is that, unlike the global method, the optimization is determined by the initial solution and the procedure ends there. This is because the territory of the initial solution that can reach the global optimum is limited. This demerit can be complemented by performing the local method multiple times, but it is complex and inefficient. Unlike the local method, the global method can solve this optimization problem efficiently by using different initial solutions.

The optimization techniques for solving combinatorial problems can be divided into the exact solution method to obtain the global optimum, and the approximate solution method to obtain a good solution that is close to the optimum. The classifications of optimization techniques are shown in Figure 5 (Kim 2017a). The exact solution methods such as linear programming for solving linear problems or branch-and-bound method and dynamic programming to solve discrete problems, are effective in various applications. However, these methods have many variables, which make it very difficult to find the optimum solution for continuous problems with high nonlinearity. Moreover, it is practically impossible to solve large combinatorial problems when the number of calculations increases exponentially with the magnitude of the problem. An alternative in this case is the heuristic method, which is an approximate method that can find an acceptable solution, instead of the global optimum, within the allowed time. The heuristic method does not guarantee the optimal solution within a reasonable calculation time but is highly likely to find a good solution. The heuristic method is mainly used to solve the representative NP-hard problem. In general, heuristic refers to a 'specific heuristic' designed to be appropriate for solving specific types of problems. Metaheuristic refers to a higher-level heuristic and a universal algorithm framework that provides guidance toward general structures and strategies for building a specific heuristic. In short, a metaheuristic is a higher-level heuristic that can be used as the basic framework for an algorithm when developing a specific heuristic (Kim 2017a).

Classification	Metaheuristics	Developer and Year of Development
Nature's evolutionary Process	Evolution Strategy(ES) Evolution Programming(EP) Genetic Algorithm(GA) Genetic Programming(GP) Differential Evolution(DE)	Schwefel(1965) Rechenberg(1965) Fogel, Owens and Walsh(1966) Holland(1975), Koza(1992) Storn & Price(1997)
Swarm	Ant Colony Optimization(ACO) Particle Swarm Optimization(PSO) Bee Colony Optimization(BCO) Artificial Bee Colony(ABC)	Dorigo(1992) Kennedy & Eberhart(1995) Lucic & Teodorovic(2001,2002) Karaboga(2005)
Natural and Social Phenomena	Simulated Annealing(SA) Tabu Search(TS) Artificial Immune System(AIS) Harmony Search(HS)	Kirkpatrick(1983) Glover(1986) Farmer et al (1986) Geem et al (2001)
Systematic Iteration	Iterated Local Search(ILS) Variable Neighborhood Search(VNS) Guided Local Search(GLS)	Baxter(1981) Mladenovic & Hansen(1997) Voudouris(1998)

Table 1: **Classification of Metaheuristics (source: Kim 2017a)**

From the mid-1960s to the present, various metaheuristics algorithms have been proposed (Beheshti and Shamsuddin 2013), and their applications are also increasing. Table 1 shows major metaheuristic methods. These metaheuristics can be categorized by the phenomena that they imitate: 1) evolution algorithm imitating the evolution process of nature; 2) algorithms imitating the behaviors of swarming organisms; 3) algorithms imitating nature and social phenomena, and 4) algorithms that searches for neighbors through systematic repetitions.

An evolutionary algorithm, which imitates the evolution of nature, including genetic algorithm, is a generic term referring to methods that use multiple evolutionary mechanisms in the design or execution of computational models for problem solving. Besides genetic algorithms, there are evolutionary strategies, evolutionary programming, and genetic programming among others. These all imitate evolutionary processes in concept through natural selection and law of inheritance. An evolutionary algorithm is not a random search technique for finding good solutions, but a stochastic process that produces better results than a random search (Kim 2017a).

While the neighborhood search techniques such as the tabu search and the simulated annealing, the algorithms imitating nature and social phenomena, operate one solution, genetic algorithms operate a population of solutions composed of multiple potential solutions. Genetic algorithms keep searching the space of solutions by applying natural selection and law of inheritance to this population of solutions from generation to generation.

Genetic algorithms are widely applied to optimization or decision-making problems because they are simple in concept and theory, and show excellent performance in searching for the number of arbitrary solutions set by researchers. In particular, genetic algorithms are appropriate for solving problems with many variables and constraints thanks to their excellent search capabilities in complex spaces of solutions. Additionally, the high flexibility of these models has the advantages of facilitating the addition of constraints and the modification of the objective function.

This study applies the genetic algorithm-based for global optimization method, which is one of the global metaheuristic methods to analyze the Optimal Train Schedule (OTS).

3.2   Genetic Algorithm

First established by (Holland 1975), the genetic algorithm is one of the representative methods that use the principle of evolution in problem-solving. Before 1975, (Rechenberg 1965) in Germany proposed the evolution strategy, followed by (Fogel, Owens, and Walsh 1966) who proposed evolution programming (Kim 2017a).

A variety of studies were conducted at similar times as Holland, but (Rechenberg 1965) used a method that transformed a single solution and did not use the crossover operation, which is the main computation method of the genetic algorithm. (Fogel et al. 1966) used mutation only without crossover operation, and this tradition is still being maintained. The framework of the group-based genetic algorithm, including crossover and mutation, was completed by (Holland 1975) (Moon 2008, Kim 2011).

Because the genetic algorithm has a simple conceptual and theoretical basis, and excellent solution search performance, it is widely applied to various optimization and decision-making problems in engineering, natural science, business administration, and social science. In particular, the genetic algorithm is appropriate for solving large mathematical problems with many variables and constraints because it has excellent search performance in a complex solution space. Other advantages of the genetic algorithm include ease of adding constraints and changing the objective function owing to the high flexibility of model application (Kim 2011).

The genetic algorithm repeats the following processes: (1) generation of initial chromosome population (2) coding of chromosomes (3) mating and mutation of chromosomes (genetic manipulation) (4) fitness evaluation (5) selection (6) generation of a new chromosome group (7) decision on whether to stop, or mating and mutation of new chromosomes, and continues searching until the best solution is found.

The first step in applying the genetic algorithm is express the potential solution of the problem as an entity. The entity expression should accurately reflect the characteristics of the problem, because it affects other processes (adaptation evaluation, etc.). Natural expressions facilitate the interpretation of entities and genetic operations.

In the train-type-based representation, a chromosome is composed of each train type instead of individual trains when the algorithm is applied. Because the actual train plan is completed by repeating the plan for each train type as many times as necessary, even if the train pattern is repeated, it has no significant effect on the solution performance. In this study, the train-type-based expression is applied to reduce the solution search time while minimizing the number of decision-making variables within a reasonable range.

In this thesis, the genetic algorithm applied to the optimal solution search process undergoes the process of the algorithm following the sequence in Figure 6.

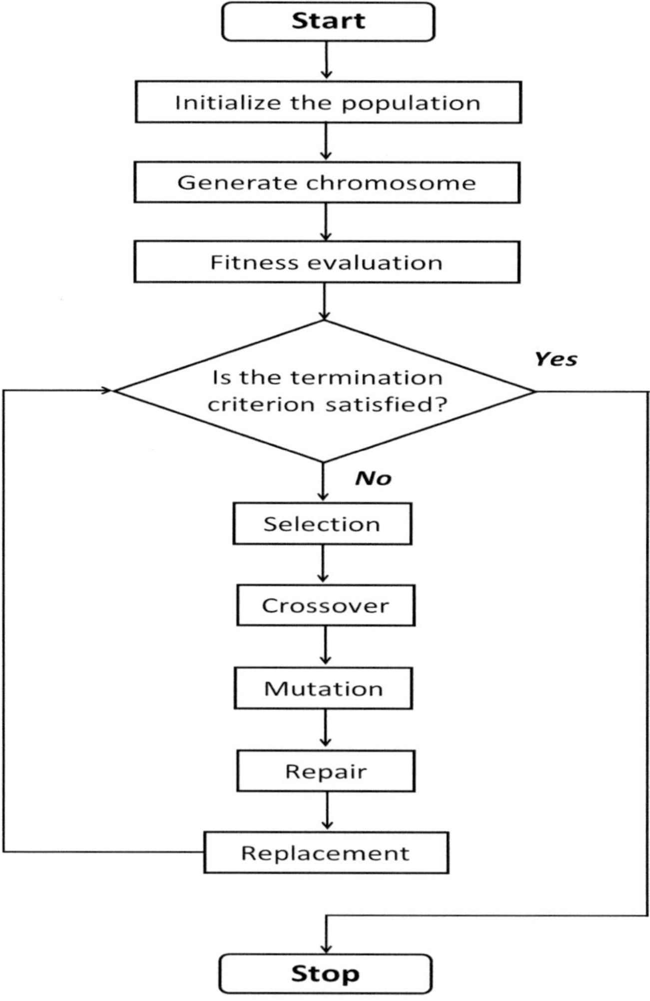

Figure 6 :   **The Flow Chart on Genetic Algorithm in This Study**

The "Initial population" is the initial step in which as many chromosomes as there are initial populations are produced. In here, a population means a set of the chromosomes and a chromosome regards as a train schedule. Considering the priority of the trains, a population is created by increasing the probability that the train with the highest priority lies at the front part of the chromosome.

In "Generate chromosome," a train schedule that satisfies all the constraints is created sequentially in accordance with the gene sequence. In this process, trains that do not meet the constraints are excluded, and the final number of trains that are finally scheduled after considering all the genes becomes the fitness evaluation of this chromosome. If the satisfaction condition is not met after this process, the algorithm moves on to selection. In this part, the chromosome to survive to the next generation is determined. The chromosomes are randomly selected and the chromosome that has a poor fitness evaluation (high scheduled waiting time) is passed to the next generation. The chromosomes are randomly selected and the chromosome that survive the selection changes superior genetic traits through crossover. After that, an operation in which random genes of the chromosome are moved to other positions through mutation is performed to maintain the diversity of solutions. In here, the gene regards as each train.

Chromosomes that have completed mutation may be inadequate. Thus, the eligibility of chromosomes is determined through a repair operation, and any ineligible chromosome is changed to an eligible chromosome. The solution that has finished the repair is replaced with one of the solutions of the current population to create a new population. In this case, the poorest quality solution is replaced and the best solution is maintained. After this operation, if the condition for satisfaction is met, then the optimal solution is derived and the operation is stopped.

In this thesis, owing to the nature of train schedule, the domain of solutions to be searched can be very large depending on the number of train operations or the number of stations, and the composition of solutions is not one-dimensional. Hence, the process to determine the eligibility and suitability of the solution is very complex, resulting in a relatively long analysis time.

4 Models for Scheduling and Delimitation

4.1 Mathematical Models for Train Scheduling

Various methods such as mathematical methods and simulation methods using optimization models are being explored to create an efficient train scheduling. High-performance computers offered computer-based scheduling methods in the 1990s (Pachl 2014) and a number of algorithms have been developed to assist making processes of train scheduling with intelligent decision support (Cacchiani et al. 2014). Furthermore, there have been many studies on modeling the Train Timetable Problem (TTP). This TTP has been proven to be an NP-Hard problem[4], which means that finding the optimal solution is impossible (Caprara et al. 2002). The existing studies approach the TTP in two parts: an objective mathematical model to find an optimal solution through a mathematical expression, and a solution method to search for the optimal solution of the mathematical model. The mathematical models to find the optimal solution can be classified into space-time, space, and location–time models, depending on the railway expression method.

Firstly, the space-time model expresses railway lines by dividing space and time into slots of a certain size. Each individual slot becomes a decision variable in the model and the number of these slots determines the size of the problem. In this case, space can be sections between stations (including intersections), or sections divided depending on the locations of railroad signals or block sections. Time can be split into seconds (1 second, 5 seconds, 10 seconds, 30 seconds, etc.) or minutes (1 minute, 5 minutes, etc.). However, as space and time are divided into smaller sized slots, the number of decision variables increases and does the scale of the problem. This may cause a case in which the problem cannot be solved within a reasonable time. Therefore, in the case of the space-time model, space and time have to be split into appropriate units (Kim 2017b).

(Lu et al. 2019) proposed a generic formulation to optimize dynamic infrastructure location and routes decisions, with a special focus on railway locomotive routing and

---

[4] NP-hardness (non-deterministic polynomial-time hardness), in computational complexity theory, is the defining property of a class of problems that are, informally, "at least as hard as the hardest problems in NP"

refueling problems using on a representation of discretized acyclic resource-space-time model, which is a combination of resource states, nodes and links of space dimension and the time dimension.

Secondly, the space model expresses railway lines by dividing railroad space into areas. (Burdett and Kozan 2010) describes railway networks as space in an extended form of the Job-shop scheduling problem. In Job shop scheduling[5], the job refers to a train and the machine refers to a portion of the track through which the train passes. However, Train scheduling problems cannot be modelled exactly as a standard job shop because of a number of features that are not commonly accommodated by the job shop scheduling perspective but are common in railways. In addition, since it is a relatively recent model, a few studies have been carried out so far and thus it can be said that the model lacks verification.

Lastly, the location-time model divides time into appropriate slots and space into locations based on the stations (including intersections) of railway lines. The number of decision variables of this model is comparatively fewer as only departure and arrival times are required to be determined for the locations. The model also represents points of railway lines (node facilities such as stations, intersections and railroad signals), which has the same expression method as the train operation diagram based on train timetables used in the actual railway work (Kim 2017b). (Tormos et al. 2008) conducted a study on the railway line between Madrid and Jaen which covers 54 locations; it is 369.4 kilometers long and has 30/23 two/one-way track sections. In this study, the location-time model and the genetic algorithm were applied. This seems to be the only case where the performance of the algorithm has been verified within a reasonable amount of time by applying the algorithm to real instances. In particular, the use of the genetic algorithm has some advantages; constraints and objective functions can be added and deleted freely; functions do not necessarily need to be linear; and objective functions and constraints have strength in the way that they can be used as long as they are expressed in mathematical formulas. As a result of comparing three models,

---

[5] Job shop scheduling is an optimization problem in computer science and operations research in which jobs are assigned to resources at particular times.

the location-time model is considered to have many advantages in terms of acceptability of various objective formulas and constraints, and extensibility and feasibility of the model. Therefore, this study uses the method of the location-time model when devising formulas for Optimal Train Schedule (OTS).

4.2 Differentiation from Existing Research

In exploring the solutions of many variable values related to train operation, it was difficult to apply genetic algorithms to the Train Timetable Problem (TTP) due to the exponentially increasing number of cases and the search space. However, improvement in computer performance and computation speed has made it possible to apply genetic algorithms to research on finding solutions to various traffic-related problems.

In this section, existing studies for train scheduling that utilized the heuristic method are discussed, and the objective of the present study, differentiated from those of existing studies, is presented.

(Wegele and Schneider 2004) proposed an algorithm for constructing timetables. The aims of this research are to simplify the control of the rail traffic process by providing an optimization algorithm, which is able to control the transport process automatically and to define optimization task as a minimization of disturbances for customer like train delays. In order to achieve the aims, this research combined constructing (branch and bound) and improving (iterative) method. The branch-and-bound method is used to construct a good start point in appropriate computational time as the first solution and an iterative improving method was applied by the genetic algorithm which has been used to improve the current solution. The method was tested with real data from German Railways.

(Gholami and Sotskov 2012) presented a genetic algorithm for routing and scheduling trains which were developed to achieve efficient and robust train routes and a timetable. This paper proposed to change the start times of the trains with the aim to find better times for the given trains to start their moving and to have a less number and a smaller total time of possible delays. As the first step, the approach of this paper is to provide a path for each train given for planning. A matrix with rows and columns corresponding to the stations was constructed. After that the algorithm was presented for finding a suitable path (train route) between the source station and the destination. This approach was tested from five trains to fifteen trains on different routes. Firstly, the problem by the data from a railway company was simulated as benchmark for computational evaluations. Then, the proposed genetic algorithm described in this paper was used to find time bill on the real data. Lastly, routing algorithm was used to change the path before scheduling in order to find a better solution of the problem. The result of the simulation was that total delay (scheduled waiting time) reduced by at least 30 minutes and by at most 130 minutes. A genetic algorithm developed and tested in this paper may help an administrator of a railway company to prepare an appropriate time bill for the set of trains.

(Nirmala and Ramprasad 2014) proposed a method of applying a genetic algorithm to calculate the optimal number of train operations considering the number of trains held. The study searched the solution of the problem through the bi-level optimization. The bi-level optimization method first determines the minimum number of train operations required for each route and sums up the number of operations by route. The study set the total number of train operations as an upper limit and proposed a method to minimize the number of train operations by utilizing the genetic algorithm considering all routes. Step 1 of the double level optimization is allocation of trains on individual routes with maximum link flow as the criteria and step 2 is further reduction of trains on network basis making use of genetic algorithms as an optimization tool.

(Arenas et al. 2015) proposed a mathematical formula that could be implemented using an optimization technique such as the meta-heuristic algorithm, and developed a model that could search an optimum solution for the assignment of the train timetable through the genetic algorithm. This paper defined the TTP that it is given a railway infrastructure, a list of available rolling stock available and a list of train journeys to schedule, produce a timetable in which, all scheduled train journeys respect both capacity and security constraints while optimizing a given criteria. Therefore, the rolling stock, the railway infrastructure and the journey requirements were identified as important inputs. The objective function is set to minimize the total travel time of general trains by optimizing stopping patterns of general trains and the number of train operations. The constraints in this paper were proposed as running time, dwell time, passing time, turnaround time and headway time. The essential components of genetic algorithm for solving the TTP in this paper represented generated initial population and a timetable. Then, the fitness function, variation operators, and combination of the current and offspring populations were used to evaluate the timetable. The study case consisted of 3 train types, and an artificial network with 15 stations, 90 station capacities (tracks inside stations) and 42 tracks, leading to 126 running and 378 headway times. Several requests sets were defined, consisting of a different number of train runs and time horizons. The results of study case using the genetic algorithm was very satisfactory. Near-optimal results were achieved within short computation times. Moreover, the model in this paper proved to be very flexible, it could be tuned to focus on other particular optimization aspect such as the reduction of total waiting time.

(Liu and Dessouky 2017) studied a decomposition of the passenger and freight train scheduling problem using heuristic algorithm. This paper considered the daily problem of scheduling both passenger and freight trains jointly. The objective is to jointly solve for the passenger and freight rail scheduling when they share the same track age to improve the efficiency of freight trains reducing their conveyance times while maintaining the punctuality of passenger trains in the same railway network. The decision variables of routing decisions, arrival/departure time and priority decision were controlled in order to optimize the objective. This research developed a heuristic algorithm that decomposes the problem into sub-problems to solve problems of realistic size because the three types of decision variables were computationally hard to solve simultaneously. The decomposition procedure of the algorithm described in this paper was that several sub-problems were solved to produce a heuristic solution to the overall problem. The paper defined three sub-problems of passenger trains, which were priority assignment, routing sub-problems and departure time. The freight trains were scheduled iteratively after the passenger train scheduling phase. At each iteration the priority and routing decisions were determined for the newly scheduled freight train. After the priority and routing decisions for this freight train is determined, all the priority and routing decisions for the passenger and the scheduled freight trains to this point are fed into a linear program to determine the time decisions for all these trains. This process is repeated until all freight trains have been scheduled.

Many studies have already been conducted to search optimal solutions by applying the genetic algorithm to train operation plans. As reviewed above with the subjects of existing studies and their main contribution, a genetic algorithm was applied to minimize of scheduled waiting time, number of trains in operation and timetable arrangement in the existing research.

Table 2 shows the differences between this study and the other studies utilizing the genetic algorithm to the optimization of train operation plans. In the case of existing studies, conveyance times vary depending on the way trains are put into service in train operation plans, which focus on the minimum train delay considering the travel time of railroad routes.

Classification	Contribution
Wegele and Schneider (2004)	The study on optimization as a minimization of disturbances for customer like train delays
Omid and Yuri (2012)	Analysis of better start times for the given trains to start their moving to have less number and a smaller total time of possible delays
Nirmala and Ramprasad (2014)	Suggestion on a method to search solutions by applying genetic algorithms to the calculation of the optimal train service frequency in consideration of the number of trains held
Arenas et al (2015)	Development of a model that allows exploration of optimal solutions for the assignment of train schedules through a genetic algorithm
Liu and Dessouky (2017)	Presentation for the passenger and freight scheduling when they share the same track to improve the efficiency of freight trains reducing their conveyance times
This study	The study on optimization of heterogeneous train scheduling structure for minimizing weighted scheduled waiting time and decision for the arrangement of the Subsidiary Main Track (SMT) for efficient train operation in railway planning.

Table 2:   Comparison between Existing Research and This Study

The study presented here analyzes the OTS based on the frequency of each train that has different characteristics, which are calculated based on traffic demand. Systematic ways should be suggested to decide combinations at an acceptable level, in order to get the optimized train scheduling in train operation for minimizing weighted scheduled waiting time, about heterogeneous trains in particular as well as the optimized placing of the track arrangement in the station in railway planning.

5 Construction of Mathematical Models for Optimization

5.1 Expression of Railway Line

The train operation plan should include calculation time, conveyance time for each train type, scheduled station-to-station running time, train formation, form and usage plan of locomotives, train timetable and train operation diagram, and use of tracks including the arrival and departure tracks within a station.

In this thesis, calculation time, number of each train type and station, scheduled station-to-station running time, maintenance of safety headways between operating trains for each formation, and prohibition of overtaking on link are considered as the constraint conditions. After the analysis of the optimal schedule, the cyclic timetable and finalized train operation diagram can be generated. A railway line can be expressed by nodes and links. The nodes include both stations and junctions, and a link is arranged between the nodes.

This thesis is used to decide the positions of the Subsidiary Main Tracks (SMTs) for overtaking in operation of heterogeneous trains, the nodes are limited to stations only. The notation used to represent the railway operation is the following.

- The nodes are expressed as $N = \{1,2,...,n\}$.

- Based on the traffic demand being planned and launched, the frequency for each train in a day is calculated. A set of the frequency for each train is expressed as $f = \{1, 2, ..., i, j, k\}$.

- All the trains that are operated in the railway lines are expressed as

  $$T = \left\{ t_1^{h,g,f}, t_2^{h,g,f}, ... t_k^{h,g,f} \right\}.$$

  where, t_k^h refers to the kth high-speed train, t_k^g to the kth general train, and t_k^f to the kth freight train. In this study, mean of a high-speed train is a train that is able to run at a speed of 200km/h or more, mean of a general train is a kind of passenger train with below 200km/h and mean of a freight train is a lower speed train than the general train.

- The number of tracks and the number of platforms in the station are expressed respectively as St_n and Sp_n, where, St_n refers to number of tracks in station and Sp_n to number of platforms in a station.

- With respect to the train operation, the arrival times of high-speed, general, and freight trains at node n are $a_n t_k^h$, $a_n t_k^g$, and $a_n t_k^f$ respectively where, $a_n t_k^h$ refers to kth high-speed train arriving at nth station, to $a_n t_k^g$ to kth general train arriving at nth station and to kth freight train arriving at nth station.

- With respect to the train operation, the departure times of high-speed, general, and freight trains at node n are $d_n t_k^h$, $d_n t_k^g$, and $d_n t_k^f$.

  where, $d_n t_k^h$ refers to kth high-speed train departing at nth station, $d_n t_k^g$ to kth general train departing at nth station and $d_n t_k^f$ to kth freight train departing at nth station respectively.

- With respect to the train operation, the headway of arrival between the trains t_k^h, t_k^g, and t_k^f at node n is expressed as follows:

$$aHw_n t_{i \to j}^{h \to h}, aHw_n t_{k \to k}^{h \to g}, aHw_n t_{k \to k}^{h \to f}$$

$$aHw_n t_{i \to j}^{g \to g}, aHw_n t_{k \to k}^{g \to h}, aHw_n t_{k \to k}^{g \to f}$$

$$aHw_n t_{i \to j}^{f \to f}, aHw_n t_{k \to k}^{f \to h}, aHw_n t_{k \to k}^{f \to g}$$

where, $aHw_n t_{i \to j}^{h \to h}$ refers to arriving headway for ith high-speed train and jth high-speed train at a station, $aHw_n t_{k \to k}^{h \to g}$ to arriving headway of for kth high-speed train and kth general train at a station and $aHw_n t_{k \to k}^{h \to f}$ to arriving headway of for kth high-speed train (h) and kth freight train at a station. As such, general trains (g) and freight trains (f) are equally applied in same sense.

- With respect to the train operation, the headway of departure between the trains t_k^h, t_k^g, and t_k^f at node n is expressed as follows.

$$dHw_n t_{i \to j}^{h \to h}, dHw_n t_{k \to k}^{h \to g}, dHw_n t_{k \to k}^{h \to f}$$

$$dHw_n t_{i \to j}^{g \to g}, dHw_n t_{k \to k}^{g \to h}, dHw_n t_{k \to k}^{g \to f}$$

$$dHw_n t_{i \to j}^{f \to f}, dHw_n t_{k \to k}^{f \to h}, dHw_n t_{k \to k}^{f \to g}$$

where, $dHw_n t^{h\to h}_{i\to j}$ refers to departing headway for i^{th} high-speed train and j^{th} high-speed train at a station, $dHw_n t^{h\to g}_{k\to k}$ to departing headway of for k^{th} high-speed train and k^{th} general train at a station and $dHw_n t^{h\to f}_{k\to k}$ to departing headway of for k^{th} high-speed train and k^{th} freight train at a station. As such, general trains (g) and freight trains (f) are equally applied in same sense.

- The operation running times, where a train $t^{h\ org\ orf}_k$ departs node n and arrives at node $n+1$, which is continuously connected, are expressed as $tt^h_k \Delta_{n,n+1}$, $tt^g_k \Delta_{n,n+1}$, and $tt^f_k \Delta_{n,n+1}$ respectively.

where, $tt^h_k \Delta_{n,n+1}$ refers to time that k^{th} high-speed train moves from a station to the next station, $tt^g_k \Delta_{n,n+1}$ to time that k^{th} general train moves from a station to the next station and $tt^f_k \Delta_{n,n+1}$ to time that k^{th} freight train moves from a station to a next station.

- When heterogeneous trains operate on the same line, the respective trains should adhere to headway between nodes. Therefore, the operation headways of the train t^h_k, t^g_k, and t^f_k must be expressed as follows.

$$\mathrm{Hw}\Delta_{n,n+1} t^{h\to h}_{i\to j},\ \mathrm{Hw}\Delta_{n,n+1} t^{h\to g}_{k\to k},\ \mathrm{Hw}\Delta_{n,n+1} t^{h\to f}_{k\to k}$$

$$\mathrm{Hw}\Delta_{n,n+1} t^{g\to g}_{i\to j},\ \mathrm{Hw}\Delta_{n,n+1} t^{g\to h}_{k\to k},\ \mathrm{Hw}\Delta_{n,n+1} t^{g\to f}_{k\to k}$$

$$\mathrm{Hw}\Delta_{n,n+1} t^{f\to f}_{i\to j},\ \mathrm{Hw}\Delta_{n,n+1} t^{f\to h}_{k\to k},\ \mathrm{Hw}\Delta_{n,n+1} t^{f\to g}_{k\to k}$$

where, $\mathrm{Hw}\Delta_{n,n+1} t^{h\to h}_{i\to j}$ refers to the headway for i^{th} high-speed train and j^{th} high-speed train at a station from a station to the next station, $\mathrm{Hw}\Delta_{n,n+1} t^{h\to g}_{k\to k}$ to headway for k^{th} high-speed train and k^{th} general train at a station from a station to the next station and $\mathrm{Hw}\Delta_{n,n+1} t^{h\to f}_{k\to k}$ to headway for k^{th} high-speed train and k^{th} freight train at a station from a station to the next station. As such, general trains (g) and freight trains (f) are equally applied in same sense.

- The safety headways of arrival for a low-speed train that is overtaken by a high-speed train at a station are expressed as $aHw_n t_{k \to k}^{f \to h}$, $aHw_n t_{k \to k}^{g \to h}$, and $aHw_n t_{k \to k}^{f \to g}$.

- The safety headways of departure for a low-speed train after overtaking a high-speed train at a station are expressed as $dHw_n t_{k \to k}^{h \to f}$, $dHw_n t_{k \to k}^{h \to g}$, and $dHw_n t_{k \to k}^{g \to f}$.

- The dwell times for the high-speed, general, and freight trains at station are expressed as $Dw_n t_k^h$, $Dw_n t_k^g$, and $Dw_n t_k^f$, respectively.

- y[s] = 1 when the SMT is installed at node n, and y[s] = 0 otherwise.

- The train priority for operation in this thesis is as follows: $t_k^h > t_k^g > t_k^f$

5.2 Mathematical Model

5.2.1 Objective Function

The objective function of the Subsidiary Main Track (SMT) installation for the optimization model is to determine the train schedule with the number of SMTs while also minimizing the total scheduled waiting time that is inevitable for the establishment of the train operation plan as a scenario given by the input data. Therefore, this thesis considers bi-objective functions.

- Conveyance time for minimization of weighted scheduled waiting time of general and freight trains caused by overtaking high-speed trains

- Determination of SMT for the overtaking high-speed train in railway track

$$min \sum_{\forall t \in T}(t \Delta \forall t + \delta_n) + \sum_{\forall t \in T}(d_n \forall t - a_n \forall t) + \sum_{n \in N} y_s \qquad \text{Formula (1)}$$

5.2.2 Constraint Function

The following constraints are necessary for efficient train schedule analysis.

- Train operation frequency during calculation time

The number of passengers or amount of freight per train must be considered when deciding on the train operation frequency. The train operation frequency is determined by the predicted traffic demand divided by transportation ability of each train.

$$Train\ operation\ frequency = \frac{Traffic\ demand}{Transportation\ ability\ of\ each\ train} \qquad \text{Formula (2)}$$

Furthermore, the number of each train per hour is calculated as the total number of operations of each train calculated on the basis of traffic demand is divided by the total calculation time. This should be considered in order to analyze efficient train schedules. In thesis, the total calculation time is considered as business hours of an operation company. For example, if the number of high-speed trains is analyzed 34 times a day and the calculation time of an operating company is assumed to be 17 hours, high-speed trains should be distributed 2 times in an hour throughout the entire business hours.

- Dwell time

The dwell time may vary by the train type and railway plan. It is a key parameter in services for users of passenger trains and for loading and unloading freights for freight trains. It is also a key parameter for train operation and planning. The dwell time can be expressed as follows:

$$Dw = \left\{ z : a_n t_k^h + dw_n t_k^h \le a_n t_k^h + dw_n t_k^h + \delta_n \right\} \qquad \text{Formula (3)}$$

where δ_n is a variability applied when the dwell time of a low-speed train which is over-taken by a high-speed train increases or when trains are overtaken or take over in a station, implying an additional dwell time.

- The guaranteed arrival of the headway of high-speed train and the guaranteed departure headway of the low-speed train

When a high-speed train overtakes a preceding low-speed train (general or freight trains), the guaranteed arrival headway of the high-speed train without reduction of speed by a preceding low-speed train and the guaranteed departure headway of the low-speed train after passing the high-speed train are as follows:

$$aHw_n t_{k \to k}^{g \to h} = \left\{ x : \left| a_n t_k^h - a_n t_k^g \right| \le Hw_n t_{k \to k}^{g \to h} \right\} \qquad \text{Formula (4)}$$

$$aHw_n t_{k \to k}^{f \to h} = \left\{ x : \left| a_n t_k^h - a_n t_k^f \right| \le Hw_n t_{k \to k}^{f \to h} \right\} \qquad \text{Formula (5)}$$

$$aHw_n t_{k \to k}^{f \to g} = \left\{ x : \left| a_n t_k^g - a_n t_k^f \right| \le Hw_n t_{k \to k}^{f \to g} \right\} \qquad \text{Formula (6)}$$

$$dHw_n t_{k \to k}^{h \to g} = \left\{ x : \left| d_n t_k^g - d_n t_k^h \right| \le Hw_n t_{k \to k}^{h \to g} \right\} \qquad \text{Formula (7)}$$

$$dHw_nt_{k \to k}^{h \to f} = \{x: |d_nt_k^f - d_nt_k^h| \leq Hw_nt_{k \to k}^{h \to f}\} \qquad \text{Formula (8)}$$

$$dHw_nt_{k \to k}^{g \to f} = \{x: |d_nt_k^f - d_nt_k^g| \leq Hw_nt_{k \to k}^{g \to f}\} \qquad \text{Formula (9)}$$

If considered the guaranteed headways are increased, it is not good results on optimal train schedule. Figure 7 is represented for the guaranteed headway between a high-speed train and a low-speed train at a station. These guaranteed headways for respective train are very important when analyzing weighted scheduled time.

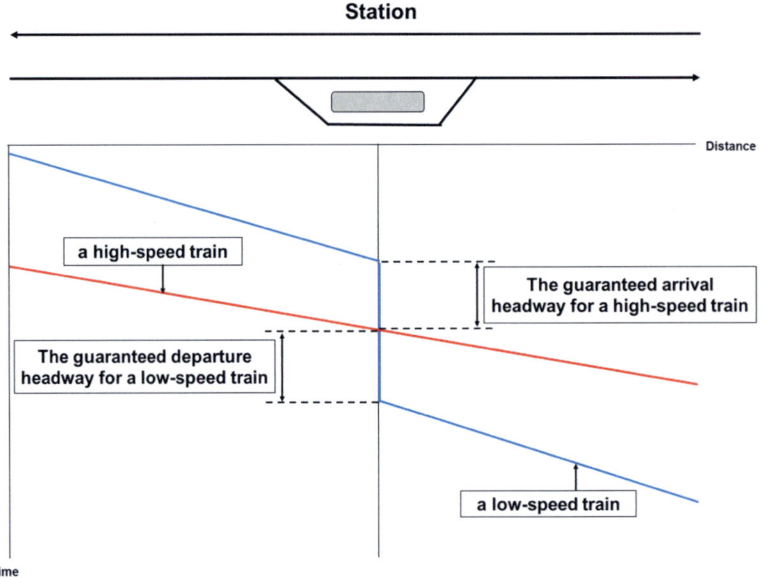

Figure 7: The Guaranteed Headway between High-speed train and Low-speed train at a Station

- Number of lines and platforms in a station

The number of trains at a specific station can be restricted if the trains stop or overtake simultaneously at a specific station, considering the lines and platforms. This is expressed as follows:

$$Meetp(t_k^p, t_k^s, n) \, if \, [a_nt_k^p, d_nt_k^p] \cap [a_nt_k^s, d_nt_k^s] \neq \emptyset \, and \, a_nt_k^{h \, or \, g \, or \, f} > 0 \; \text{Formula (10)}$$

$$meet(t_k^p, t_k^s, n) \, if \, [a_nt_k^p, d_nt_k^p] \cap [a_nt_k^s, d_nt_k^s] \neq \emptyset \, and \, a_nt_k^{h \, or \, g \, or \, f} = 0 \quad \text{Formula (11)}$$

$$Meetp(t_k^p, t_k^s, n) < Sp_n \, and \, Meetp(t_k^p, t_k^s, n) + meet(t_k^p, t_k^s, n) < Sl_n \quad \text{Formula (12)}$$

In other words, $[a_n t_k^p, d_n t_k^p] \cap [a_n t_k^s, d_n t_k^s] \neq \emptyset$ implies that there is a train at node n. And then, $a_n t_k^{h \, or \, g \, or \, f} > 0$ signifies that one of a high-speed train or a general train or a freight train arrives at same node n for more than 0 minute. In this case, it is expressed as $Meetp(t_k^p, t_k^s, n)$. In addition, $a_n t_k^{h \, or \, g \, or \, f} = 0$ represents that one of a high-speed train or a general train or a freight train passes at node n. Therefore, $meet(t_k^p, t_k^s, n)$ represents that one of a high-speed train or a general train or a freight train passes at node n when there is a train at node n. The number of trains occupying a yard line or platform at node n is counted; from this, the satisfaction of the physical environment constraints can be determined. Figure 8 is represented for this constraint.

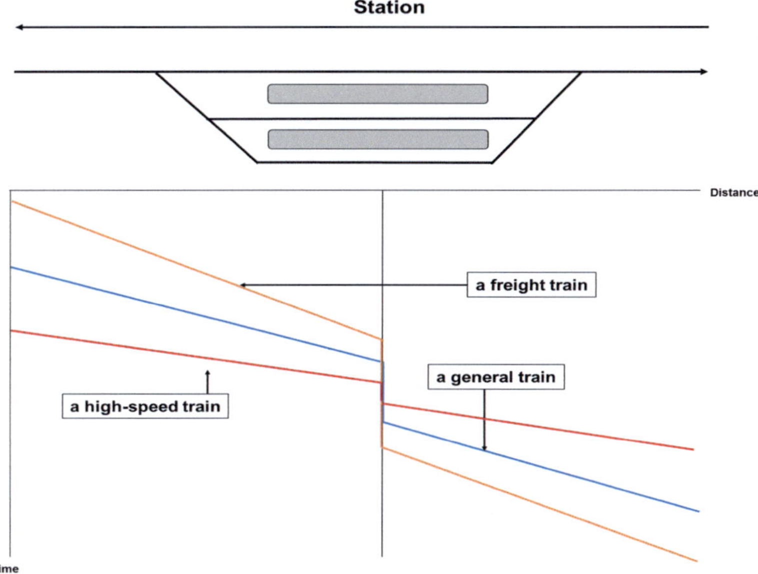

Figure 8:   **The Case of Stopping Trains Simultaneously at a Station**

- Conditions for non-occurrence of overtaking between stations (link)

This thesis assumed that the sidetrack or middle track that can be used for the overtaking of trains is not in a line between stations. Thus, the condition for not overtaking between trains in a line between stations is expressed as follows:

$$(d_{n-1} t_k^p < d_{n-1} t_k^s) \implies (a_n t_k^p < a_n t_k^s) \qquad \text{Formula (13)}$$

- Running time

The running time of a train is determined by the arrival time of the next station from a station at which the starting time is fixed.

$$tt_k^h \Delta_{n,n+1} = a_{n+1}t_k^h - d_n t_k^h \qquad \text{Formula (14)}$$

$$tt_k^g \Delta_{n,n+1} = a_{n+1}t_k^f - d_n t_k^g \qquad \text{Formula (15)}$$

$$tt_k^f \Delta_{n,n+1} = a_{n+1}t_k^f - d_n t_k^f \qquad \text{Formula (16)}$$

- Determination of a station at which to install a SMT for overtaking

In this thesis, under the assumption that the distance between stations is fixed, when high-speed and low-speed trains run on the same tracks, the low-speed trains are overtaken by the high-speed trains and the high-speed trains overtake the low-speed trains. In this case, the condition for determining a station that requires SMTs for overtaking is expressed as follows.

$$(d_n t_k^g + t_k^g \Delta_{n,n+1} + aHw_{n+1}t_{k\to k}^{g\to h} \wedge a_{n+1}t_k^g - t_k^g \Delta_{n,n+1} - aHw_{n+1}t_{k\to k}^{g\to h}) < (d_n t_k^h +$$
$$Hw_n t_{k\to k}^{g\to h} + t_k^h \Delta_{n,n+1} \wedge a_{n+1}t_k^h - Hw_n t_{k\to k}^{g\to h} - t_k^h \Delta_{n,n+1}) \qquad \text{Formula (17)}$$

$$(d_n t_k^f + t_k^h \Delta_{n,n+1} + aHw_{n+1}t_{k\to k}^{f\to h} \wedge a_{n+1}t_k^f - t_k^f \Delta_{n,n+1} - aHw_{n+1}t_{k\to k}^{f\to h}) < (d_n t_k^h +$$
$$Hw_n t_{k\to k}^{f\to h} + t_k^h \Delta_{n,n+1} \wedge a_{n+1}t_k^h - Hw_n t_{k\to k}^{f\to h} - t_k^h \Delta_{n,n+1}) \qquad \text{Formula (18)}$$

$$(d_n t_k^f + t_k^f \Delta_{n,n+1} + aHw_{n+1}t_{k\to k}^{f\to g} \wedge a_{n+1}t_k^f - t_k^f \Delta_{n,n+1} - aHw_{n+1}t_{k\to k}^{f\to g}) < (d_n t_k^g +$$
$$Hw_n t_{k\to k}^{f\to g} + t_k^g \Delta_{n,n+1} \wedge a_{n+1}t_k^g - Hw_n t_{k\to k}^{f\to g} - t_k^g \Delta_{n,n+1}) \qquad \text{Formula (19)}$$

This is necessary for the optimization of train schedules as well as the optimization of facilities, and a decisive factor for line capacity analysis. This thesis proposes an optimization model for a SMT installation to be applied to complex train operation scenarios. The SMT installation optimization model proposed in this study calculates the optimal number of SMTs using the train scheduling model.

- Blocking Time

Blocking time (German: "Sperrzeit") is the total elapse time a section of track (e.g. a block section, an interlocked route) is allocated exclusively to a train movement and therefore blocked for other trains (Pachl 2014; Cao 2017). The blocking times directly determine the signal headway as the minimum headway between two successive trains, considering only one block section. Therefore, it is possible to determine the minimum headway between two trains (Pachl 2014).

The blocking time finishes after the train has completely left the section and all signaling appliances have been reset to normal position so that movement authority can be issued to another train to enter the same section. Therefore, the blocking time of a track section is usually much longer than the time the train occupies the section. As shown in Figure 9, the blocking time of a block section for a train without a scheduled stop at the entrance of that section consists of the following time intervals in a territory with fixed block signal system (Pachl 2014).

- the time for clearing the signal
- a certain time for the driver to view the clear aspect at the signal in rear that gives the approach indicator to the signal at the entrance of the block section
- the approach time between the signal that provides the approach indicator and the signal at the entrance of the block section
- the time between the block signals, as running time
- the clearing time to clear the block section and – if required- the overlap with the full length of the train
- the release time to "unlock" the block system

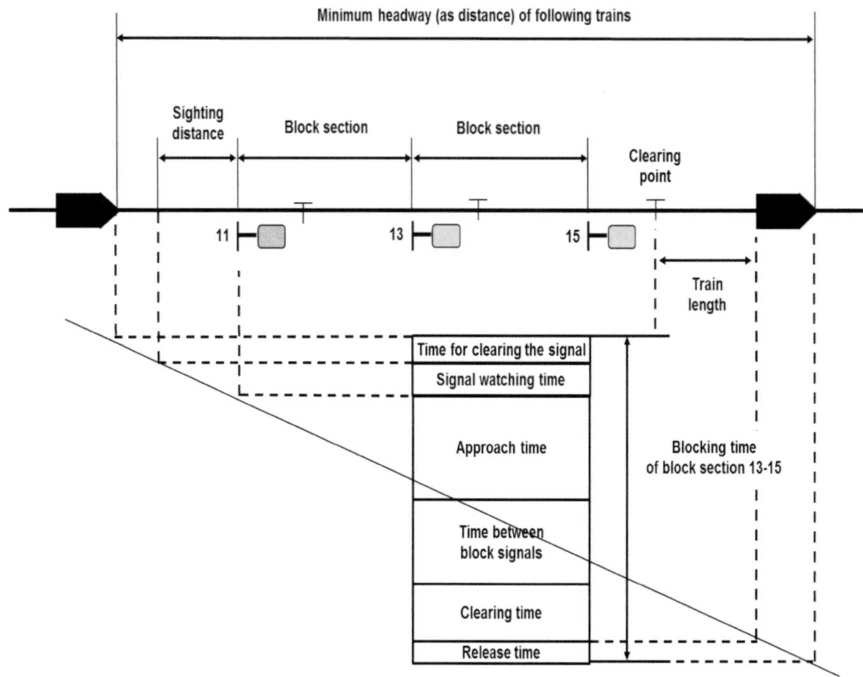

Figure 9:   **Blocking time of a Block Section (source: Pachl 2014)**

The minimum headway is a key parameter in selecting the SMTs that can be used when high-speed trains pass low-speed trains. In this study, the analysis of the optimum train scheduling determines the stations that require SMTs for overtaking of high-speed trains. To analyze the minimum headway between trains, a line with the longest distance between stations is selected in a station with SMTs from which a low-speed train has started. For the following high-speed train to run at a non-braking speed, the time when the preceding low-speed train arrives at the station with SMTs and the guaranteed minimum headway to be overtaken by following high-speed are required. To meet this condition, the headway Hw between the fastest high-speed train t_k^h and the lowest speed train is required for the low-speed train t_k^f, which is expressed as follows:

$$Hw = \left\{ x : x \in ebtt_{k,bt}^f - bbtt_{k,bt}^h \right\} \text{ for bt} = 1 \ldots \ldots n_b \qquad \text{Formula (20)}$$

6 Optimization Algorithm Modeling

The purpose of the optimization algorithm for heterogeneous train schedules in this thesis is to implement a train schedule with minimized weighted scheduled waiting time due to low-speed trains (general and freight trains) into a Subsidiary Main Track (SMT) for overtaking by high-speed trains while operating the trains a number of times for each type within the specified calculation time; considering all the constraints for the three types of trains (high-speed, general, and freight); and determining the locations of the stations that require overtaking, that is, the location of the SMT.

The Optimal Train Schedule (OTS) can have multiple goals, but in this thesis, it is to minimizing the scheduled waiting time caused by preceding low-speed trains by following high-speed trains, which is necessitated by the competition between high-speed and low-speed trains. The scheduled waiting time is minimized by considering the constraints mentioned in Chapter 5, and a weight is assigned to the delay by train type. The train priority for operation in this thesis is given in the order of high-speed train, general train, freight train. The weights are applied to give high-speed trains to top priority and to set priorities between general trains and freight trains when implementing algorithms. In other words, high-speed trains can pass general trains and freight trains as they have priority over general and freight trains. General trains can only overtake freight trains and freight trains are required to allow passing high-speed trains and general trains for their operation. Therefore, while scheduled waiting time is not caused by other trains during operation of high-speed trains, it occurs in operation of general trains due to high-speed trains, and in operation of freight trains due to high-speed trains and general trains. Freight trains may overtake general trains in actual train operations. However, this study does not take into account the cases where low-speed trains are overtaken by high-speed trains. The weights can be adjusted in further studies, if such cases should be considered.

Fitness is defined as the sum of the scheduled waiting time considering the weights with respect to the train type for a specific train schedule. The trains are scheduled to have a fitness lower than the target fitness. It can be added to the goal simply by changing the train schedule generation code and adding it to the constraints or changing the fitness evaluation code, because major operations deal with the location of the SMT in the stations for purposes other than train schedule fitness evaluation. Therefore, a lower fitness means a minimization of weighted scheduled waiting time and the actual fitness is as follows in this thesis.

$$\text{Fitness} = \sum_{\text{order of general train}=1}^{\text{number of general trains}} \text{the scheduled waiting time of general train}$$
$$\times \text{ the weight of general train}$$
$$+ \sum_{\text{order of freight train}=1}^{\text{number of freight trains}} \text{the scheduled waiting time of freight train}$$
$$\times \text{ the weight of freight train}$$

Formula (21)

6.1   Methods and Procedures

For optimal train scheduling, this thesis is based on a genetic algorithm, which is a global optimization method that finds the optimal solution by searching the domain of all solutions.

The process for determining the eligibility and suitability of the solution is very complex, resulting in a relatively long calculation time due to the nature of train scheduling. Therefore, several considerations and some local optimization methods are added to narrow the domain of solutions to be searched.

The basic process is as follows:

1. Generate random train schedules.
2. Evaluate the fitness of train schedules.
3. Generate new train schedules that more fit (or highly likely to be more fit) by considering the fitness of the current train schedules.
4. Repeat steps 2 and 3 until a train schedule with a fitness lower than the target fitness is generated.

When a new train schedule is generated in step 3, the genetic algorithm may require different times to find a solution that meets the same goal under the same conditions, depending on how much the diversity or convergence of solutions are considered. In this thesis, the algorithm is implemented so that it is easy to input and change various parameters, such as calculation time, number of operations by train type, and stations.

The data input and output are based on hypertext markup language 5(HTML5)[6], the graph output on scalable vector graphics (SVG)[7], and the algorithm implementation on Javascript/jQuery[8]. These were configured for research, but the user interface could be modified for more universal utilization. If the algorithm was converted into a server-based algorithm, then the results could be obtained under more complex conditions with shorter response time

6.2   Data Structure

The major data structures include the train schedule, detailed operation schedule, overtaking and scheduled waiting time, and station data.

6.2.1   Train Schedule Data

The train schedule data structure is a list (array) of train types and departure times for individual operation. It is the shape of a solution obtained from the OTS and corresponds to the chromosome in the genetic algorithm. Thus, the train schedule comprising the train types and departure times become the genes that comprise a chromosome.

Train schedule = chromosome = solution

There are three types of trains, i.e., high-speed, general, and freight trains, and the departure time is the time when the train leaves the first station counted from the operation start time of zero.

The detailed operation schedule data have the following structures:

---

[6] HTML5 is the latest standard for the basic programming language HTML, which is used to produce web documents.

[7] SVG is an open-source vector graphic file format based on Extensible Markup Language (XML) for expressing two-dimensional vector graphics. It was developed under the initiative of the World Wide Web Consortium (W3C) in 1999. XML was created to overcome the limitations of HTML by facilitating data exchange between heterogeneous systems connected to the Internet.

[8] JavaScript is a prototype-based script programming language mainly used in Web browsers, and has a feature to access objects embedded in other application programs

- A freight train that departs at 1 minute
- A general train that departs at 14 minutes
- A freight train that departs at 18 minutes
- .....
- A high-speed train that departs at 171 minutes
- A freight train that departs at 176 minutes

It can be expressed in pseudo code[9] as follows.

train schedule = [[freight train, 1], [general train, 14], [freight train, 18], ... [high-speed train, 171], [freight train, 176]]

Once the operation starts, detailed train schedule with the scheduled waiting time can be added.

6.2.2   Detailed Operation Schedule

The detailed operation schedule data structure is a list (array) of the arrival times for all stations excluding the first station, and the departure times for all stations excluding the last station for one train operation, which is added to each train operation in the train schedule.

The detailed train schedule includes the scheduled waiting time caused by overtaking high-speed trains to prevent railway line overlapping by trains departing at different times. The overtaking and the data of scheduled waiting time become the basis of the fitness evaluation in the genetic algorithm. Furthermore, the fitness of a solution can be finally confirmed by calculating the detailed operation schedule.

It is possible to determine the fitness to some degree in advance with other methods, but the detailed operation schedule must be produced to accurately determine the fitness. If a solution is not fit, then the departure time itself must be changed for overtaking; if the departure time is changed, then the current solution becomes a different

---

[9] Pseudo code is an informal high-level description of the operating principle of a computer program or other algorithm. It uses the structural conventions of a normal programming language, but is intended for human reading rather than machine reading

solution that no is longer fits. The arrival and departure times of a station are the times (minutes) counted from the operation start time of zero.

The detailed operation schedule data have the following structures:

- Departure time from first station +1 minute
- Arrival time at second station +28 minutes
- Departure time from second station +64 minutes
- …
- Arrival time at last station +271 minutes

In pseudo code, they are expressed as follows.

detailed operation schedule = [1, 28, 64, …, 271]

6.2.3   Scheduled waiting time

The scheduled waiting times is a list (array) of train types that caused an overtaking, the departure time of the train that caused an overtaking, and the overtaking station that has a SMT for an overtaking. This is added when the detailed operation schedule is produced.

The scheduled waiting time is zero if a train departing at another time passes through the station during the dwell time at the stopping station, but a SMT is required. Thus, the scheduled waiting time is considered in this case as well.

The scheduled waiting time becomes the criteria for evaluating fitness in the genetic algorithm, and the location of the SMT can be confirmed in the corresponding train schedule. The stations are defined sequentially from 0 when creating the algorithm model. Therefore, this study represented the first station as 0 and the second station as 1 in pseudo code.

The scheduled waiting time has the following structures:

- Departure delayed by 10 minutes from the second station owing to a general train departing at +49 minutes
- Departure delayed by 13 minutes from the third station owing to a high-speed train departing at +71 minutes

- ...

- Departure delayed by 0 minutes from the fourth station owing to a general station departing at +95 minutes

In pseudo code, they are expressed as follows.

```
scheduled waiting time = [[49, general train, 1, 10], [71,
high-speed train, 1, 13],…, [95, general train, 3, 0]]
```

The structure with the detailed train schedule and the scheduled waiting time are as follows.

```
train schedule = [

    [

        freight train,

        1,

        [1, 28, 64, … 271],

        [[49, general train, 1, 10], [71, high-speed train, 1, 13], [95,
    general train, 3, 0], …]

    ],

    …

]
```

6.2.4   Station

The station data structure is a list (array) of travel times from the previous to the present station, and stop/non-stop by train type. The station data becomes the basic data for performing various operations such as the calculation of the detailed operation schedule. The first station (departing station) is omitted because it has no previous station and stop/non-stop has no meaning in this algorithm. The stop/non-stop at the last station is also omitted.

The station data have the following structures:

- Running time from the first to second station, high-speed train 10 minutes, general train 18 minutes, freight train 27 minutes; at the second station, high-speed train stop, general train stop, freight train non-stop
- Running time from the second to the third station, high-speed train 15 minutes, general train 26 minutes, freight train 40 minutes; at the third station, high-speed train non-stop, general train stop, freight train non-stop
- …
- Running time from the sixth to the last station, high-speed train 10 minutes, general train 18 minutes, freight train 27 minutes

6.3 Structure of Genetic Algorithm on Optimization of Train Schedule

The structure of the genetic algorithm implemented in this study to analyze the OTS is as follows.

```
start the genetic algorithm = function() {

    population = [];

    for (number of chromosomes in one generation) {

        population.push (generate a random train schedule ());

    }

    repeat  = true;

    while(repeat) {

      for (order 1; population) {

        for (order 2; train schedule) {

          train schedule[order 2] = calculation of detailed operation schedule(train schedule[order 2]);

        }

        population[order].fitness = fitness evaluation (population[order 1]);
```

```
            if (population[order].fitness < target fitness) number of fit solutions
   ++;

      }

   if (number of fit solutions >= 3) repeat = false;

   else {

     for (i = 0; populatin chromosome count -4; i ++) {

       selected chromosome = select chromosome ();

       chromosome = crossover (selected chromosome);

       chromosome = mutation (chromosome);

       chromosome = repair (chromosome);

       population = sort in ascending order of fitness (population);

       population[last].delete();

       population.push(chromosome);

      }

     }

    }

                        }
```

6.3.1   Creation of Population

A random train schedule that meets the given conditions, such as calculation time and train operation frequency by train type, is created. In the genetic algorithm, the initial population is determined by repeating this process for the number of chromosomes of the population by a generation that requires this process.

6.3.1.1   Generation of Arbitrary Train Schedule

When a train schedule (chromosome) is generated, the following conditions must be met.

- Trains depart as many times as the operation count per train type within the operating hours.
- The calculation time and operation count by train type are input values. Passenger trains cannot depart at an interval shorter than the minimum interval or longer than the maximum interval.
- Passenger trains mean high-speed trains and general trains.
- Passenger train minimum interval (minute) = Passenger train expected interval - passenger train adjustable time x 2
- Passenger train maximum interval (minute) = Passenger train expected interval + passenger train adjustable time x 2
- Passenger train expected interval (minute) = Calculation time (minute) / number of passenger trains
- The passenger train adjustable time (minute) is an input variable.
- The shorter the passenger train adjustable time, the closer to the expected interval the passenger train departs, but it is more likely that it will not become the OTS.
- Every train must depart at an interval longer than the minimum operational time interval. The minimum operational time interval (minute) including buffer time is an input variable.

The process for generating a random train schedule (chromosome) that meets the above conditions is as follows.

1. A list (array) of departure times that accurately maintains the expected passenger train interval within the calculation time is generated.

   e.g.) Departure times for passenger trains if the passenger train operation count at an calculation time of 180 minutes is 9 =

   [10, 30, 50, 70, 90, 110, 130, 150, 170]

```
list of departure times for passenger trains = [];

generate a list of departure times for passenger trains = function() {

    first passenger train departure time = passenger train expected interval / 2;

    for (i = First passenger train departure time; i <= calculation time; i += passenger train expected interval) {
```

```
        list of departure times for passenger trains.push(i);

    }

};
```

2. A passenger train schedule is created by first randomly allocating high-speed trains for the operation count, and then allocating general trains for the remainder, considering the minimum and maximum intervals for high-speed trains with the array of departure times produced in Case 1. The minimum and maximum intervals for high-speed trains are required to obtain more useful results by evenly distributing the departure times of two types of passenger trains, which are determined as follows.

- Minimum intervals of high-speed train(minute) = operation time(minute) / number of high-speed train x 0.75

- Maximum intervals of high-speed train(minute) = operation time(minute) / number of high-speed train x 1.25

    e.g.) When high-speed trains and general trains are each 3 and 6 times, passenger train schedule is as follows: [[high-speed train, 10], [general train, 30], [general train, 50], [high-speed train, 70], [general train, 90], [general train, 110], [high-speed train, 130], [general train, 150], [general train, 170]]

```
generate passenger train schedule = function() {

    passenger train schedule = []

    arrangement of high-speed train positions = []

    minimum positions = 0;

    maximum positions = (high-speed train maximum interval – first passenger train departure time) / expected interval of passenger train);

    random position = generate random positions (minimum position, maximum position);

    arrangement of high-speed train positions.push (random position);

    for (i = 0 ; i < (number of high-speed trains - 1); i ++) {
```

```
        minimum positions = random positions + high-speed train minimum in-
terval / passenger train expected interval;

        maximum positions = random positions + high-speed train maximum
interval / passenger train expected interval;

        random positions = generate random positions (minimum position,
maximum position);

        arrangement of high-speed train positions.push (random positions);

    }

    for (list of departure times for passenger trains) {

        train type = "general train";

        for (arrangement of high-speed train positions) {

            if (departure time for passenger trains == position of high-speed
train) {

                train type = "high-speed train";

                break;

            }

        };

        passenger train schedule.push([train type, departure time for passen-
ger trains]);

    };

    };
```

3. The passenger train schedule generated in number 2 is randomly adjusted
   within the passenger train adjustable time (minute). When adjusting the passenger train schedule, the adjusted departure time may not be fit in some cases.
   Readjust it if the departure time is unfit; if readjustment to a fit departure time
   is impossible, then classify it as an exception and stop.

e.g.) When the passenger train adjustable time is ± 5, then the passenger train schedule after adjustment is as follows: [[high-speed train, 7], [general train, 34], [general train, 52], [high-speed train, 74], [general train, 89], [general train, 110], [high-speed train, 126], [general train, 155], [general train, 172]]

```
for(passenger train schedule) {
    passenger train departure time = random adjustment of passenger
train departure time (passenger train departure time);
  });
  random adjustment of passenger train departure time = (departure time)
{
    minimum = - passenger train adjustable time/sampling interval;
    maximum = passenger train adjustable time/sampling interval;
    repeat or not = true;
    number of repetitions = 0;
    while (repeat or not) {
        randomly adjusted departure time = departure time + random gen-
eration (minimum, maximum) * sampling interval;
        exception or not = check for exception(randomly adjusted departure
time);
        number of repetitions
        if (!exception) {
            repeat or not = false
        }
      if (number of repetitions == 1000) {
          repeat or not
          log(number of repetitions + "random adjustment of passenger train
departure time failed within");
        }
      }
      return randomly adjusted departure time;

    };
```

4. A list (array) of departure times for which freight train operations can be added to the passenger train schedule generated in number 3 is produced. This list excludes the departure times that generate unfit solutions such as departure times with an interval longer than the minimum calculation time interval for a passenger train departure time, and departure times that must be changed owing to delays or other reasons. If a list of sufficient freight train departure times for the number of freight trains cannot be obtained depending on the calculation time, number of operations by type, or other conditions, then classify it as an exception and stop.

```
Add freight train schedule = function(train schedule) {

    for (i = 0; i < number of freight trains; i ++) {

        list of addable train count and departure times = Check if freight trains
    can be added (train schedule);

        if (addable train count < freight train count) {

            exception();

            break;

        }

        minimum =  0;

        maximum = list of addable departure times - 1;

        selected departure time = List of addable departure times [Generate
    random number (minimum, maximum)];

            train schedule.push([freight train, Selected departure time]);

            train schedule = sort train schedule (train schedule);

    }

    return train schedule;

};
```

6.3.1.2 Limitations of the search area of solution to reduce incidence of ineligibility solutions

Unfit solutions can occur frequently if a train schedule is generated randomly without limiting the search domain of solutions, in addition to the given conditions. This limitation should be applied when generating a train schedule because it can take much more time if an unfit solution is found after a detailed operation schedule, which is then discarded to generate a new train schedule.

In this thesis, if the intervals between high-speed, general, and freight trains in the train schedule are those with a very high likelihood of generating unfit solutions derived by analyzing the station data, then the corresponding departure times are changed to reduce the probability of unfit solutions.

```
for(order; station list) {

    high-speed train arrival time by station.push (high-speed train departure
time by station[order - 1] + high-speed train travel time by station[order]);

    general train arrival time by station.push(general train departure time by sta-
tion[order - 1] + general train travel time by station[order]);

    freight train arrival time by station.push(freight train departure time by sta-
tion[order - 1] + freight train travel time by station[order]);

    high-speed and general trains arrival time difference by station.push(general
train arrival time by station[order] - high-speed train arrival time by station[or-
der]);

    high-speed and freight trains arrival time difference by station.push(freight
train arrival time by station[order] - high-speed train arrival time by station[or-
der]);

    general and freight trains arrival time difference by station.push(freight train
arrival time by station[order] - general train arrival time by station[order]);

};

check available departure times for freight train considering high-speed train =
function(list of available departure time sections, train schedule) {
```

```
while(repeat) {

  number of repetitions ++;

  repeat = false;

   for(order 1; List of available departure time sections) {

    $.for(order 2; train schedule) {

       if (train schedule[order 2] == high-speed train && train schedule[order
2].departure time -freight and high-speed trains minimum interval > available
departure start time  && train schedule[order 2].departure time - freight train
calculation time with no delay < available departure end time) {

          for(order3; High-speed and general trains arrival time difference by
station) {

             minimum time = train schedule[order 2].departure time - high-speed
and general trains arrival time difference by station[order3] - safe arrival time;

             maximum time = train schedule[order 2].departure time - high-speed
and general trains arrival time difference by station[order3];

             if (minimum time < available departure start time  && maximum
time >= available departure end time) {

                List of available departure time sections.delete(order 1);

                repeat = true;

                break;

             }

             if (minimum time >= available departure start time && minimum time
<= available departure end time - sampling interval) {

                list of available departure time sections.delete(order 1);

                if (maximum time <= available departure end time - sampling inter-
val) {
```

```
            list of available departure time sections.push (available departure
start time, minimum time);

            list of available departure time sections.push (maximum time +
sampling interval, available departure end time);

        } else {

            list of available departure time sections.push(available departure
start time, minimum time);

        }

        repeat = true;

        break;

}

        if (minimum time < available departure start time  && maximum
time >= available departure start time ) {

        list of available departure time sections.delete(order 1);

        if (available departure end time >= maximum time + sampling inter-
val) {

            list of available departure time sections.push(maximum time +
sampling interval, available departure end time);

        }
        repeat = true;

                break;

        }

    });

    }

    if (Repeat) {

      break;
```

```
      }
    });

    if (Repeat) {

      break;

    }

   };

  }

  return list of available departure time sections;

 };

 Check available departure time for freight train considering general train =
 function (list of available departure time sections, train schedule) {

    ...

 };

 Check available departure time for general train considering high-speed train
 = function (list of available departure time sections, train schedule) {

    ...

 };
```

The time intervals with a very high likelihood of unfit solutions include the basic case in which the departure time must be changed because overtaking is required before arriving at the first station, and the case in which the departure time must be changed because sufficient time for safe arrival cannot be guaranteed when waiting for another high-speed train is required at any station, unless many changes occur due to external factors.

This method can contribute to reducing the time required to find the optimal solution. However, if the low-speed trains must be overtaken in several times, then it could have a low probability of being a fit solution with a low probability. Because overtaking for a relatively long period of time must be made at a certain station, even if it can become a fit solution, it is less probable that it will become the optimal solution; there is also a possibility that a candidate for the optimal solution is excluded.

If more time can be allotted to find the optimal solution, then the random train schedule generation code can simply be modified to omit the exception handling part.

6.3.2　Fitness Evaluation

Here, fitness is the sum of the scheduled waiting time considering the weight by train type. To verify the scheduled waiting time of each solution, that is, each train schedule comprising the current population, the detailed operation schedule is produced and the fitness is determined and evaluated through the scheduled waiting time generated at this time.

6.3.2.1　Calculation of Detailed Operation Schedule

To produce a detailed operation schedule, a train schedule and station data are required. Station data are input variables.

When producing a detailed operation schedule, the following criteria must be met.

1. The velocity performance of each train is listed in the order of high-speed trains, general trains and freight trains. If there are infrastructure restrictions, each train should comply with the restriction.
2. For a section where a high-speed train that departed later is faster than a general/freight train that departed earlier, if the section passes through a station or if it includes a station where the train stops before departing again, then the general/freight train is overtaken at the station. If the section is between two stations, then the general/freight train is overtaken at the previous station and waits for the proper departure time after the high-speed train has overtaken it before departing again.
3. For a section where a general train that departed later is faster than a freight train that departed earlier, if the section passes through a station or if it includes a station where the train stops before departing again, then the freight train is overtaken at the station. If the section is between two stations, then the freight train is overtaken at the previous station and waits for the proper departure time after the general train has overtaken it before departing again.

If a general/freight train that departed later overlaps with a train of the same type that departed earlier, then the train that departed later is overtaken and waits for the minimum calculation time interval after the train that departed earlier departs before departing again. High-speed trains are excluded because they do not delayed due to low-speed trains, and thus there are not overlap with each other in this study.

The production process of a detailed operation schedule is as follows.

1. The initial detailed train schedule data are created by calculating every section time under the assumption that there are no delays in departure time for every train operation in the train schedule.

```
for (train schedule) {

    detailed operation schedule = [];

    if (high-speed train) {

      for(high-speed train section time) {

        detailed operation schedule.push (departure time + high-speed train section time);

      };

    } else if (general train) {

        for(general train section time) {

detailed operation schedule.push (departure time + general train section time);

        };

      } else if (freight train) {

        for(freight train section time) {

          detailed operation schedule.push (departure time +
    freight train section time);

        };

      }

        current train.push(detailed operation schedule);
```

```
};
```

2. Based on the detailed operation schedule data, determine if there is a case where a high-speed train that departed later arrives earlier than a general train that departed earlier by comparing the general trains that departed earlier than the high-speed trains, in the order of the earliest to the latest departure time. If such a case exists, then modify the detailed operation schedule data of the general train in accordance with the calculation criteria for the detailed operation schedule, and generate the scheduled waiting time for the corresponding general train. Check the scheduled waiting time and exclude cases where overtaking between two trains already occurred.

```
for(order 1; train schedule) {

  if (order 1 high-speed train) {

    for(order 2; train schedule) {

      if (order 2 general train) {

        for(scheduled waiting time) {

          if(order 1 high-speed train == overtaking high-speed train) {

            overtake = true;

            break;

          }

        }

  if (!overtake && (general train departure time < high-speed train departure time
  && general train arrival time + safe arrival spare time > high-speed train arrival
  time)) {

            scheduled waiting time is required = true;

            train to overtake = order 1 high-speed train;
```

```
                train to be overtaken = order 2 general train;

            end = true;

            break;

        }

        if (order 2 == (order 1 - 1)) {

            end = true;

            break;

        }

        }

    });

    }

    if (end) break;

};
if (scheduled waiting time is required) adjust detailed operation schedule for
overtaking (train to overtake, train to be overtaken);
```

3. Based on the detailed operation schedule data, determine if there is a case where a general train that departed later arrives earlier than a freight train that departed earlier by comparing the freight trains that departed earlier than the general trains, in the order of the earliest to the latest departure time. If such a case exists, then modify the detailed operation schedule data of the freight train in accordance with the calculation criteria for the detailed operation schedule, and produce the scheduled waiting time for the corresponding freight train. Check the scheduled waiting time and exclude cases where overtaking between two trains already occurred.

```
for(order 1; train schedule) {

    if (order 1 high-speed train || general train) {
```

```
    for(order 2; train schedule) {

  if (order 2 freight train) {

    for(scheduled waiting time) {

      if(order 1 train == overtaking train) {

        overtake = true;

        break;

      }

    }

      if (!overtake && (freight train departure time < high-speed/general train
departure time && freight train arrival time + safe arrival spare time > high-speed
/general train arrival time)) {

        scheduled waiting time is required = true;

        train to overtake = order 1 high-speed /general train;

        train to be overtaken = order 2 freight train;

        end = true;

        break;

      }

    }

  });

  }

  if (order 2 == (order 1 - 1)) {

      end = true;

      break;

    }

  }
```

```
    });

  }

  if (end) break;

};

if (scheduled waiting time is required) adjust the detailed operation schedule
for overtaking (train to overtake, train to be overtaken);
```

4. Based on the detailed operation schedule data, determine if there is a case where a general/freight train that departed later overlaps with a train of the same type that departed earlier by comparing the general/freight trains of the same type that departed earlier than the corresponding general/freight train, in the order of the earliest to the latest departure time. If such a case exists, then modify the detailed operation schedule data of the general/freight train that departed later in accordance with the calculation criteria for the detailed operation schedule, and generate the scheduled waiting time for the corresponding general/freight train. In this case, even if there has been an overtaking between two trains, it is not excluded.

```
for (order 1; train schedule) {

  if (order 1 general train) {

    for(order 2; train schedule) {

      if (order 2 general train) {

if (order 2 general train departure time < order 1 general train departure time

&& order 2 general train arrival time + minimum calculation time interval >

order 1

general train arrival time) {

        (scheduled waiting time is required) = true;

        overlapping train = order 2 general train;

        train to be overtaken = order 1 general train;

        end = true;

        break;

      }

      if (order 2 == (order 1 - 1)) {

        end = true;

        break;

if (end) break;

};

if (scheduled waiting time is required) Adjust detailed operation schedule

without overtaking (overlapping train, train to be overtaken);

  for(order 1; train schedule) {

    if (order 1 freight train) {

      for(order 2; train schedule) {

        if (order 2 freight train) {
```

```
        if (order 2 freight train departure time < order 1 freight train departure
time && order 2 freight train arrival time + minimum calculation time interval >
order 1 freight train arrival time) {

        (scheduled waiting time is required) = true;

        overlapping train = order 2 freight train;

        train to be overtaken = order 1 freight train;

        end = true;

        break;

    }

    if (order 2 == (order 1 - 1)) {

        end = true;

        break;

if (end) break;

if (scheduled waiting time is required) adjust detailed operation schedule
without overtaking (overlapping train, train to be overtaken);
```

5. If the scheduled waiting time take places at a specific station because of the occurrence of number 2 to 4, then delete the station data from the existing scheduled waiting time, and restore the part of the detailed train schedule data from which the scheduled waiting time has been deleted.

6. Repeat processes 2 to 5 until number 2 to 4 no longer occur.

6.3.2.2   Calculation of Fitness

The fitness is calculated through the scheduled waiting time in the overtaking generated when producing the detailed train schedule. Before calculating and evaluating fitness, the following conditions must be established:

- There must be a general train delay weight and freight train delay weight. These two values are input by researchers.
- There must be a target fitness. Target fitness is used to finish the process when a solution with a fitness lower than the target fitness is obtained.

```
scheduled waiting time of general train = 0;
scheduled waiting time of freight train = 0;
for(train schedule) {
   if (general train) {
   scheduled waiting time of general train += general train arrival time -
general train departure time – running time of general train without delay;
   }
  if (freight train) {
      scheduled waiting time of freight train += freight train arrival time -
freight train departure time – running time of freight train without delay;
      }
   };
 fitness = scheduled waiting time of general train * weights for scheduled
waiting time of general train + scheduled waiting time of freight train *
weights for scheduled waiting time of freight train;
```

The fitness is the sum of scheduled waiting time of all general trains in the train schedule, multiplied by the general train delay weight plus the sum of scheduled waiting time of all freight trains, multiplied by the freight train delay weight. Furthermore, this fitness is equal to the sum of the arrival time differences between the initial detailed train schedule for which the scheduled waiting time has not been confirmed, and the final detailed train schedule for which the scheduled waiting time have been confirmed, multiplied by the weight. After calculating the fitness, it is compared with the target fitness. If there are solutions with fitness lower than the target fitness (in this case, if there are three or more solutions), then the corresponding solutions are set as the optimal solutions and the process is completed. Otherwise, a new population is created in accordance with the genetic algorithm.

6.3.3   Generation of New Population

If the target solution is not found in the current population after the fitness evaluation, then a new population is created based on the fitness of the current population. When creating a new population, the goal is to create a population that is more fit (or likely to be fit).

For the new population, select two chromosomes (train schedules) from the current population, create new chromosomes by crossing the genes (train operations), induce mutation with a certain probability to obtain a diversity of solutions, and determine whether the newly-created train schedule is fit. If not, then repair it. Finally, if the new solution is fit, then replace it with the solution of the current population to create a new population. After the new population is created, the fitness evaluation is performed again.

6.3.3.1   Selection

Select two train schedules for crossover using the quality-proportional roulette wheel method in this case. The quality-proportional roulette wheel method (Kim 2017a) involves randomly selecting two solutions after adjusting the probability such that the probability of selecting the fittest train schedule is k times higher than that of the least fit train schedule.

A higher k value can increase the probability of selecting a solution with better quality (minimization of weighted scheduled waiting time), but it can also decrease the diversity of solutions. In this thesis, the k value is set to 3.for (order; population) {

```
    if(order == 0) {

        maximum fitness = population[order].fitness;

        best fitness = population[order].fitness;

    }

    if(population[order].fitness > maximum fitness) maximum fitness = population[order].fitness;
```

```
    if(population[order].fitness < best fitness) best fitness = population[order].fit-
ness;

  }

  for(order; population) {

    population [order].start roulette wheel area = (order == 0) ? 0 : population[or-
der - 1].stop roulette wheel area - 1;

    population[order].stop roulette wheel area = best fitness * 4 - (best fitness * 2
* (population[order].fitness - best fitness) / (maximum fitness – best fitness) +
best fitness);

  }

  random number 1 = generate random number(0, population[last].stop roulette
wheel area);

  random number 2 = generate random number(0, population[last].stop roulette
wheel area);

  for(order; population) {

    if (random number 1 > population[order].start roulette wheel area && random
number 1 < population[order].stop roulette wheel area) chromosome 1 = popu-
lation[order];

    if (random number 2 > population[order].start roulette wheel area && random
number 2 < population[order].stop roulette wheel area) chromosome 2 = popu-
lation[order];

  }
```

6.3.3.2   Crossover

A new train schedule is created through the two train schedules selected above.

The equal-crossing method is used by default. However, the data are sorted by train type and departure time before crossing them, because the train types at the crossed position may be different.

The equal crossing method (Kim 2017a) is used to generate a random number at each gene position and create a new chromosome by using the genes of chromosome 1 if the random number is less than the specified value, and using the genes of chromosome 2 otherwise. Here, chromosomes are created by generating a random number between 0 and 1, and distinguishing them based on 0.5.

```
new chromosome = [];
  chromosome 1 = sort by train type (chromosome 1);
  chromosome 2 = sort by train type (chromosome 2);
  for(order; chromosome 1) {
    random number = generate random number(0, 1);
    if (random number < 0.5) new chromosome.push (chromosome 1[order]);
    else new chromosome.push (chromosome 2[order]);
}
```

6.3.3.3   Mutation

Mutation means transforming some genes in the chromosome from Section 6.3.3.2 with a certain probability to ensure the diversity of solutions. Here, a typical mutation is used, where some of the genes in the chromosome are randomly changed with a probability of 1.5%.

```
Random number = generate random number(0, 1000);
if (random number <= 15) {
  random number = generate random number(0, chromosome length - 1);
  chromosome[random number] = get a new departure time ();
}
```

6.3.3.4   Repair

The chromosome (train schedule) that has completed until mutation is likely to become an unfit solution. Therefore, the fitness of a solution is determined; if it is unfit, then it is changed to a fit solution.

```
for(order; train schedule) {

    train schedule[order] = calculation of detailed operation schedule(train
schedule[order]);

    if(!train schedule[order].fitness) {

      train schedule.delete(order);

      train schedule[order] = get a new departure time ();

    }

}
```

6.3.3.5 Replacement

The solution that has completed the repair process is replaced with one solution of the current population to create a new population. The best three solutions are maintained and the poorest-quality solution is replaced.

```
population = sort in descending order of fitness(population);

population[last].delete();

population.push(chromosome);
```

6.4 The Unified Modeling Language (UML)

The Unified Modeling Language (UML) is a graphical purposed-oriented modeling language for visualizing, specifying, constructing and documenting the artifacts of a software-intensive system (Cao 2017). The types of diagrams in UML are as follows: a use case diagram that demonstrates the interactions among the elements of a system; an activity diagram that models the work flow or shows the life cycle of the object; a sequence diagram that shows object interactions arranged in time sequence; a collaboration diagram, also called a dynamic diagram, which shows the sequence of messages exchanged among the objects during the interaction; a class diagram that describes the structure of a system; a component diagram that describes the physical or logical components in a system; and a deployment diagram that shows the physical architecture of a system and the relationships between the software and hardware components in the system (Choi 2018).

6.4.1   Class Diagram

The class diagram shows the responsibilities that each class has to offer. Most classes have responsibilities for managing and maintaining certain data, and for performing some tasks such as computation or input or output. A class refers to a small module that defines the data to be processed by the system and the operation associated with the data. The data of the class have data values for the algorithm of this study (Choi 2018).

This study uses class diagrams to present the structure of the genetic algorithm. The class diagram is used to well describe connections, generalizations, and collective re-lationships among classes. It also effectively demonstrates the functions, especially attitudes and operations, and the interface of a class. In addition, it shows what objects can exist within the algorithm.

The elements of the class diagram are shown in Table 3. Classes are represented by boxes, as shown in Table 3. Divide the box into three parts, and write the class name in the top blank. Write the attributes of the class in the middle blank and the operations and parameters that use the data in the bottom blank. Attributes refer to the data of the elements of the algorithm model, which is abstracted into a class. For example, the elements of delivery have the information of the order, destination, and contact number, as well as the expected delivery date as their attributes. The operation displays the actions or functions that the class can execute and lists the type and the name of the parameter in parentheses. The lines among classes show their connection. The connection demonstrates the relations of the underlying classes serving as lubricant in the class diagram. The connection has to be well marked so that no problem occurs in implementing the functions of the system in the future. While finding a class is important, finding relations among classes is essential for understanding its meaning correctly.

Notation	Name	Contents
name attribute operation()	Class	Display the things that have to be extracted from the algorithm and stored. Enter the class name in the top blank, the attribute in the middle, and the operation in the bottom.
attribute type	Attribute	Save the data showing the state of the object.
operation (name & type)	Operation	Show actions or functions that the class can execute and write the type and the name of the parameter in parentheses.
————	Relation	Display the relations between a numbers of classes. Mark the name of the role on both sides to detail the relations. Mark the number of objects participating in the connection, i.e. the multiplicity.
◀ – – –	Dependency	A dependency is a relationship that signifies that a single or a set of model elements requires other model elements for their specification or implementation.
◆————	Composite relation	Relations that include the objects of overall and partial concepts and share the object's fortunes.

Table 3: **The elements of the Class Diagram in this Study (source: Choi 2018; Cao 2017)**

6.4.2   Class Diagram for Structuring the Genetic Algorithm

The chapter 6.3 in this thesis represents the class diagram as Figure 10. To create a class diagram, it should be identified which a class is in a system. Then, it is necessary to verify a relation between an attribute and the class in the system. The class, which is the building block of class diagrams, contains data that represent attributes and operations that represent actions.

The Figure 10 shows the train schedule that includes trains, train routes, and the solution (train schedule), which is the set of solutions of genetic algorithms. Create an initial set of solutions using the chromosome object in the chromosomes class, and then make train schedules. Adjust the departure time of the high-speed and the general trains and calculate the departure time of freight trains accordingly. Next, find the solution that has the lowest fitness through selection, crossover, mutation, repair, and replacement. The evaluation class evaluates the set of solutions and creates data for the scheduled waiting time. The limitation class creates train schedules by limiting the unconformable solutions, and the guaranteed time is readjusted based on the evaluation result of the train schedule. Repeat these steps until the target fitness given by the researcher is met, and analyze the shortest scheduled waiting time, namely the chromosome with the best fitness. It is considered as the optimum train schedule within the target fitness in this study. The some elements of class diagram in Figure 10 are coded using the following abbreviations like Table 4.

Elements	Abbreviation	Elements	Abbreviation
Departure Time	DT	Arrival Time	AT
Passenger Train	PT	General Train	GT
High-speed Train	HT	Freight Train	FT
Guaranteed Arrival	GA	Guaranteed Departure	GD
Passenger Train Schedule	PTS	Freight Train Schedule	FTS

Table 4: **The abbreviations for the some elements**

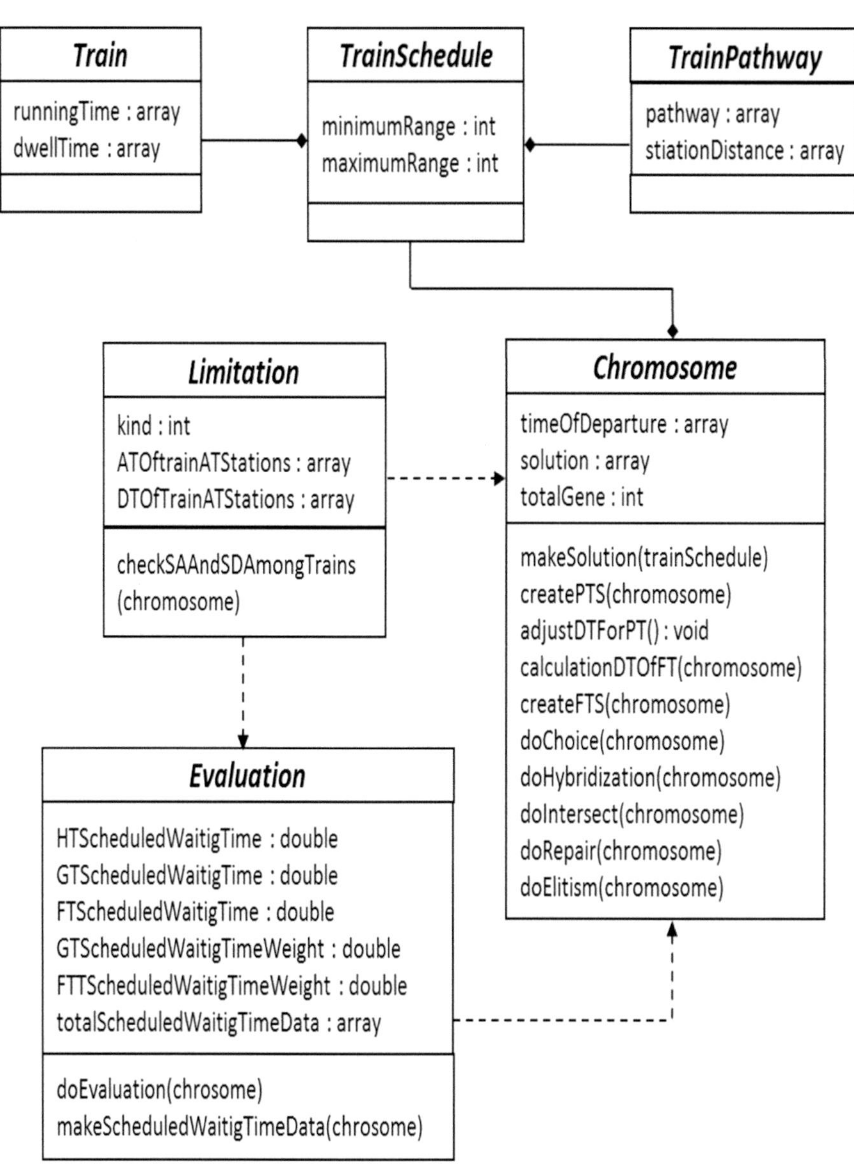

Figure 10:   **Class Diagram for the Structure of Genetic Algorithm**

7 Case Study Using the Optimization Algorithm

7.1 Applied Railway Line

The railway line to which the mathematical model developed above and the genetic algorithm will be applied serves the function of connecting the western region of the capital area and the Honam region in the South–North direction in Republic of Korea. It encompasses the high-speed train function to prepare for the future inter-Korea rail transportation. In the short term, this line will solve the problem of insufficient line capacity of the Seoul–Busan axis by supplying high-speed trains to the Chungcheong region and providing spatiotemporal railway services. In the long term, this line will provide the main railway functions in preparation for the connection of the inter-Korean railway. The railway planning process is currently in the strategic phase.

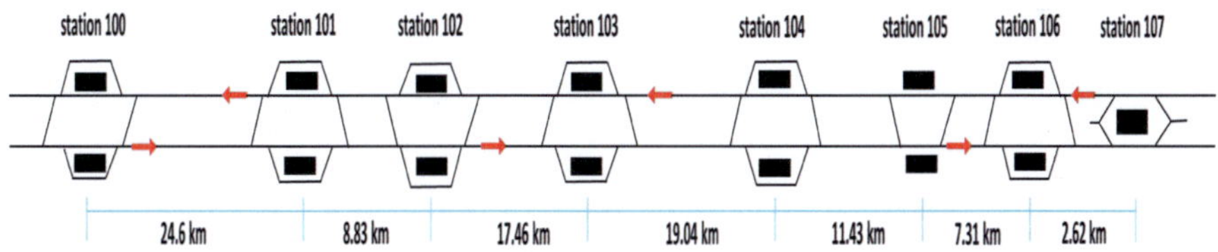

Figure 11:   **Layout of Stations for Applied Railway Line**

The layout of stations for applied railway line is shown in Figure 11. All station between the Station 100 and the Station 107 have a Subsidiary Main Track (SMT) except for the Station 105. When analyzing this railway line it is assumed that there is no a SMT at all stations in order to compare current layout of stations with analysis result used the genetic algorithm in this study described.

Section	Station 100 to 107	
Maximum Velocity	V = 250km/h	
Maximum gradient	G = 12.5 ‰	
Minimum radius	R = 1,200m	
Effective length of a station	600m and over	
Type of train	Passenger	High-speed train, General train
	Freight	Freight train

Table 5:   **Railway Line Standards and Type of Trains**

Table 5 represents the standard of the railway line and train types on this line. The maximum velocity of this line for operation is 250km/h and train types are passenger train as high-speed train and general train, and freight train. They operate at different speed, dwell time, location of stop station and operation frequency of each train in this line. Input values must be set to apply this model, such as the section distance of the applied railway line, operation frequency of each train during calculation hours, running time between stations, and location of stop stations. General trains stop for 1 minutes at all middle stations excluding the departure station 100 and the arrival station 107, high-speed trains stop for 2 minutes at Station 103, and freight trains stop for 20 minutes at Station 103 and 106. Therefore, the total dwell time of each train shall be 6 minutes for general trains, 2 minutes for high- speed trains, and 40 minutes for freight trains.

Table 6 represents parameters that are considered as input values. The running time of each train was analyzed using the Train Performance Simulation (TPS) of the Korea Rail Network Authority. TPS is a train simulation tool for testing the train performance of a track. It was developed for accurate and safe performance of simulated running based on actual running to test a new line or a line design for improvement. Running TPS requires caution, and the performance of the input vehicles, various speed limits, and train operation regulations must be strictly applied. In addition, the section running time in the train plan that can guarantee practicality is derived and utilized as basic data. The running time between stations was calculated and used in the plan through the TPS based on theoretical and dynamic of train movement conditions such as distance, speed, actual running time, and energy consumption during train operation as well as general factors such as line, vehicle, and operation conditions. However, the section running time must be finalized through trial operation. The standard running time was calculated as shown in Table 6. The minimum unit of the standard running time is 15 seconds. The train operation frequency (number of train service) for a day is calculated by considering the transportation ability of each train based on the traffic demand analyzed by the Korea Transport Database. Railway companies in Republic of Korea operate railway service for about 18 hours a day. Therefore, the calculation time of the train schedule in this study is 18 hours. In the calculation time, therefore, the train operation frequency for each train considered in this thesis is represented in Table 6.

Division		Distance	Freight Train (f)		General Train (g)		High-speed Train (h)	
Station		Total 91.29km	Running time be-tween stations (minute)	Velocity between stations (km/h)	Running time be-tween stations (minute)	Velocity between stations (km/h)	Running time be-tween stations (minute)	Velocity between stations (km/h)
100	101	24.60	17.25	85.87	10.75	137.30	7.25	203.6
101	102	8.83	6.00	88.31	4.25	124.68	2.25	235.5
102	103	17.46	12.25	85.50	7.75	135.15	5.25	199.5
103	104	19.04	13.75	83.08	8.50	134.39	6.00	190.4
104	105	11.43	7.75	88.49	5.50	124.69	2.75	249.4
105	106	7.31	6.00	73.10	3.75	116.96	2.25	194.9
106	107	2.62	3.25	48.33	2.00	78.54	1.75	89.8
Total running time[10]			66.25 minutes		42.50 minutes		27.50 minutes	
Total dwell time			40.00 minutes		6.00 minutes		2.00 minutes	
Transport time[11]			106.25 minutes		48.50 minutes		29.50 minutes	
Velocity			51.55 km/h		112.93 km/h		185.67 km/h	
Train operation frequency[12]			15 numbers		19 numbers		22 numbers	

Table 6:   **Parameters of Each Train and Train Operation Frequency**

---

[10] Total running time is train running time of the whole railway line excluding the dwell time.

[11] Transport time is the total of all the running and dwell times in a train's end to end journey (train related) or total hours during which the service is operated (passenger train related).

[12] Train operation frequency is number of each train for a day.

7.2   Evaluation in Accordance with Target Fitness

The methodology used in this paper to analyze the optimal train schedule (OTS) is a genetic algorithm. The algorithm examines random train schedules (chromosomes) to find schedules that have less than the target fitness (scheduled waiting time). To find the most OTS, the same target fitness is repeated several times to find the result values, and the OTS (optimal solution), which has the best fitness points (scheduled waiting time), is analyzed. To find the OTS (optimal solution) of the route being used, 100 randomly analyzed train schedules (chromosomes) are examined to find the train schedule with the lowest fitness, i.e. the train schedule with the minimization of weighted scheduled waiting time for low-speed trains being passed by high-speed train operations. As explained in Chapter 3, the genetic algorithm repeats the following tasks to find the optimal solution.

One cycle of this task is called 1 generation, and 300 cycles of this task are called 300 generations. In this study, the above task is set to be performed 300 times to find the OTS (nearly optimal solution). Ultimately, the train schedules (chromosomes) are analyzed for 300 generations, and this task is performed 100 times. The train schedule (chromosome) with the minimization of weighted scheduled waiting time out of 100 train schedules (chromosomes) is the OTS (nearly optimal solution). Four target fitness of 70, 65, 60 and 55 are analyzed to compare cases where the fitness becomes lower by 5-minutes increments.

7.2.1   Target Fitness of 70

In Figure 12, each point shows the fitness point observed through the task of repeating the genetic algorithm 300 times, as well as the generation which found the fitness point. Most fitness points were between 64 and 70, and most fitness points were observed in less than 50 generations. The fitness point of 47.75 weighted scheduled waiting time observed in the 4th generation is from the OTS with the minimization of weighted scheduled waiting time (red point) out of 100 train schedules below the target fitness of 70.

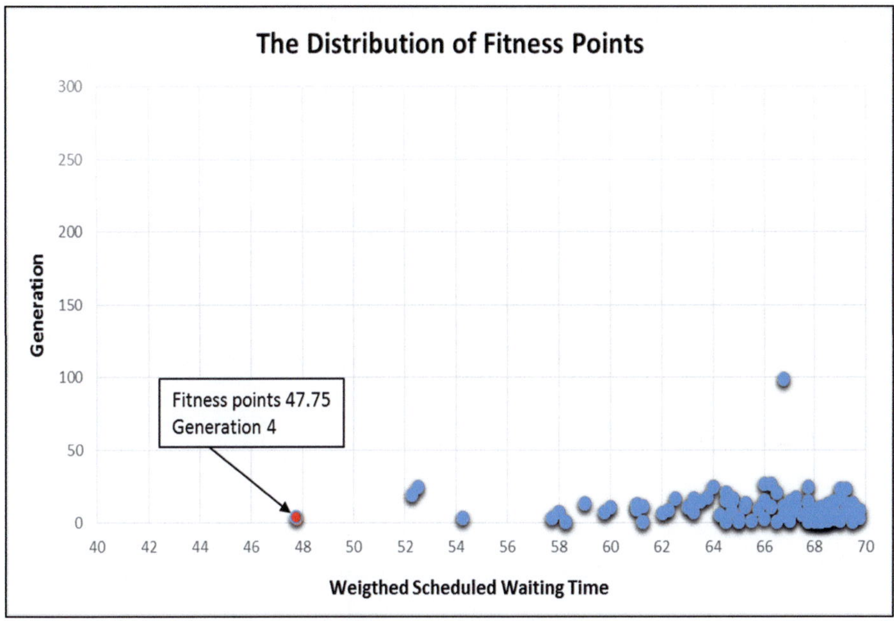

Figure 12: **Distribution Chart of Fitness Points for Target Fitness 70**

During 18 hours of operation, low-speed trains were overtaken by high-speed trains at all middle stations, including 101, 102, 103, 104, 105, and 106, but excluding the departure station (100) and the terminal station (107). In the optimal schedule at the target fitness of 70, there were no instances that high-speed trains and general trains are overtaken freight trains or high-speed trains are overtaken by general trains.

Overtaking Station	Number of Train Schedules (chromosomes)
101, 102, 103, 104, 105, 106	97
101, 102, 103, 105, 106	2
101, 102, 103, 104, 106	1

Table 7: **Number of Train Schedules where the Same Overtaking Station in Target Fitness of 70**

Table 7 shows the number of train schedules where the same overtaking station was requested out of the 100 train schedules. There were 97 train schedules where an overtaking occurred at Stations 101, 102, 103, 104, 105, and 106. There were two train schedules where an overtaking occurred at Stations 101, 102, 103, 105, and 106. There was a train schedule where an overtaking occurred at Stations 101, 102, 103, 104 and 106.

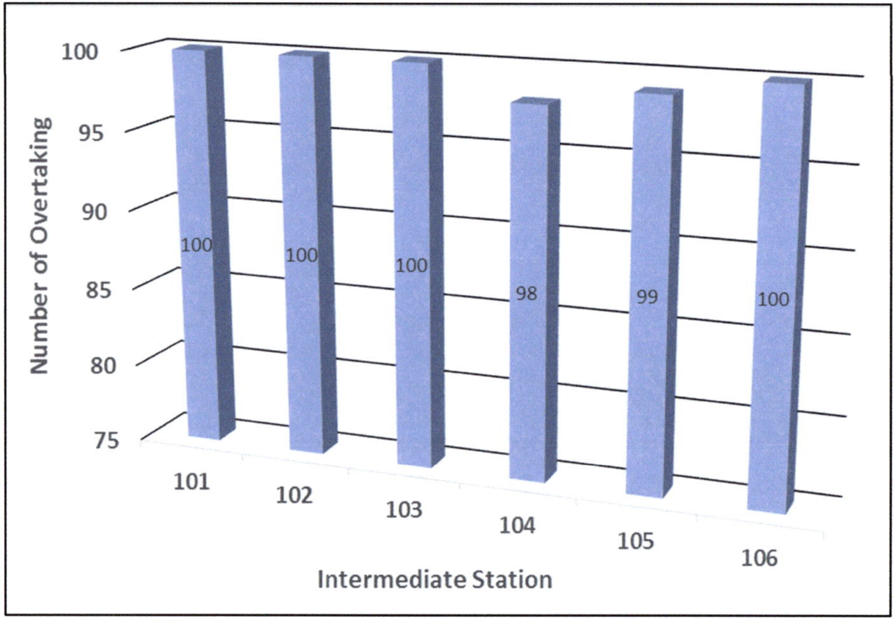

Figure 13: **Number of Overtaking on Each Station for Target Fitness of 70**

Figure 13 shows the number of times low-speed trains were overtaken by high-speed trains even once at each station in the 100 analyzed train schedules. Station 105 was observed to have the least overtaking at 99 times, followed by Station 104 at 98 times. The overtaken trains occurred in all 100 train schedules at Stations 101, 102, 103 and 106.

Table 8 shows the fitness of train schedules with five or less overtaking stations among six intermediate stations (101,102,103,104,105,106) excluding the departing and the terminating stations, the generation that finds this fitness, and the stations where Subsidiary Main Tracks (SMTs) are required at the target fitness of 70.

No	Generation	Fitness	Station where Subsidiary Main Tracks (SMTs) are required for overtaking
1	9	62.25	101,102,103,105,106
2	1	68.25	
3	13	67	101,102,103,104,106

Table 8:　　　**Fitness Point for 5 Overtaking Stations in Target Fitness 70**

It is analyzed that the two, among the three train schedules, do not require SMTs at Station 104, and the other does not require SMTs at Station 105. The lowest fitness of the train schedule with five or less overtaking stations at the target fitness of 70 is 62.25 weighted scheduled waiting time, which is derived at the 9th generation, and the concerned schedule does not require auxiliary main lines at Station 104. The train schedule with a best fitness of 47.75 weighted scheduled waiting time derived from the 4th generation at the target fitness of 70 requires SMTs at all the intermediate stations, therefore, the difference in fitness of these train schedules is 14.5 weighted scheduled waiting time.

7.2.2　　Target Fitness of 65

In Figure 14, each point shows the fitness point observed through the task of repeating the genetic algorithm 300 times, as well as the generation which found the fitness point. Most fitness points were between 58 and 65, and most fitness points were observed in less than 100 generations. The fitness point of 47.75 weighted scheduled waiting time observed in the 29th generation is from the OTS the minimization of weighted scheduled waiting time (red point) out of 100 train schedules below the target fitness of 65.

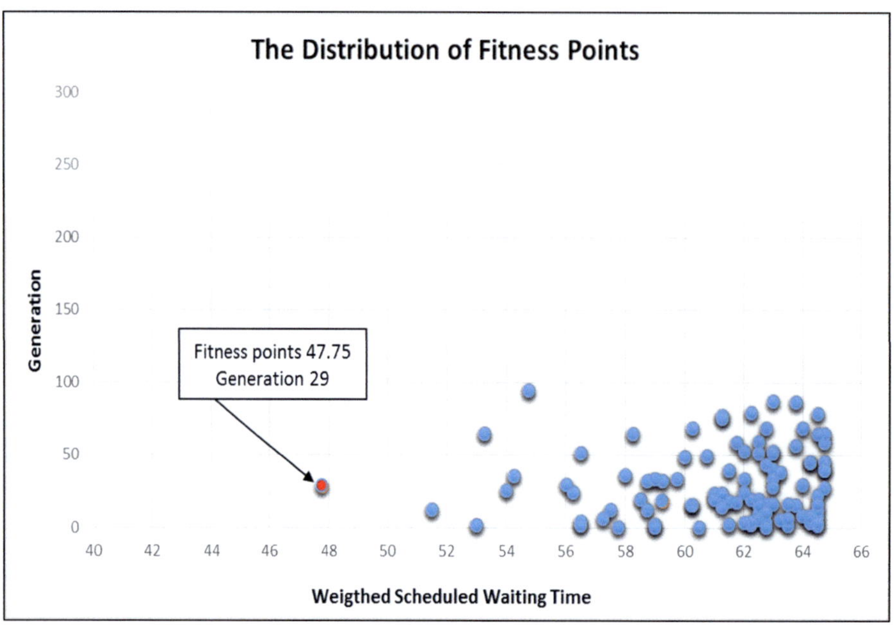

Figure 14:   **Distribution Chart of Fitness Points for Target Fitness 65**

During 18 hours of operation, low-speed trains were overtaken by high-speed trains at all middle stations, including 101, 102, 103, 104, 105, and 106, but excluding the departure station (100) and the terminal station (107). In the optimal schedule at the target fitness of 65, there were no instances that high-speed trains and general trains are overtaken by freight trains or high-speed trains are overtaken by general trains.

Overtaking Station	Number of Train Schedules (chromosomes)
101, 102, 103, 104, 105, 106	79
101, 102, 103, 104, 106	14
101, 102, 103, 105, 106	7

Table 9:   **Number of Train Schedules where the Same Overtaking Station in Target Fitness of 65**

Table 9 shows the number of train schedules where the same overtaking station was requested out of the 100 train schedules. There were 79 train schedules where overtaking occurred at Stations 101, 102, 103, 104, 105, and 106. There were fourteen train schedules where an overtaking occurred at Stations 101, 102, 103, 104, and 106. There were seven train schedules where overtaking occurred at Stations 101, 102, 103, 105, and 106.

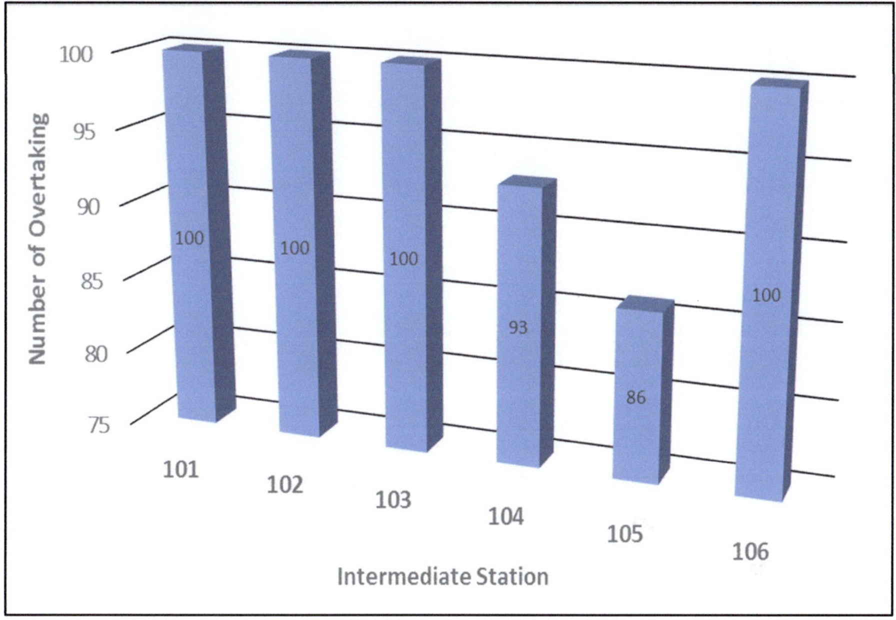

Figure 15: **Number of Overtaking on each Station for Target Fitness 65**

Figure 15 shows the number of times low-speed trains were overtaken by high-speed trains even once at each station in the 100 analyzed train schedules. Station 105 was observed to have the least overtaking at 86 times, followed by station 104 at 93 times. Overtaking of trains occurred in all 100 train schedules at Stations 101, 102, 103, and 106.

Table 10 shows the finesses of train schedules with five or less overtaking stations among six intermediate stations (101,102,103,104,105,106) except for the departing and the terminating stations, the generations that find these fitness, and the stations where SMTs are required at the target fitness of 65 weighted scheduled waiting time.

	Generation	Fitness	Station where Subsidiary Main Tracks (SMTs) are required for overtaking
1	2	53	101,102,103,105,106
2	78	62.15	
3	16	63.75	
4	37	58	
5	20	61	
6	23	64.5	
7	13	57.5	
8	53	62	101,102,103,104,106
9	87	63.75	
10	17	63.5	
11	57	63.75	
12	30	56	
13	49	60	
14	17	62.75	
15	7	57.25	
16	95	54.75	
17	16	62.5	
18	2	64.5	
19	52	62.5	
20	29	63	
21	20	58.5	

Table 10:   **Fitness Point for 5 Overtaking Stations in Target Fitness of 65**

It is analyzed that seven train schedules do not require SMTs at Station 104, and 14 train schedules do not require SMTs at Station 105. The lowest level of fitness for train schedules with five or less overtaking stations at the target fitness of 65 is 53 weighted scheduled waiting time, which is derived at the 2th generation, and the concerned schedule does not require SMTs at Station 104.

The train schedule with the best fitness of 47.75 weighted scheduled waiting time, which is derived at the 29th generation at the target fitness of 65, requires SMTs at all the intermediate stations, therefore, the difference in fitness among these train schedules is 5.25 weighted scheduled waiting time.

7.2.3   Target Fitness of 60

In Figure 16, each point shows the fitness point observed through the task of repeating the genetic algorithm 300 times, as well as the generation which found the fitness point. Most fitness points were between 54 and 60, and most fitness points were observed in less than 300 generations. The fitness point of 47.25 weighted scheduled waiting time observed in the 72th generation is from the OTS with the minimization of weighted scheduled waiting time (red point) out of 100 train schedules below the target fitness of 60.

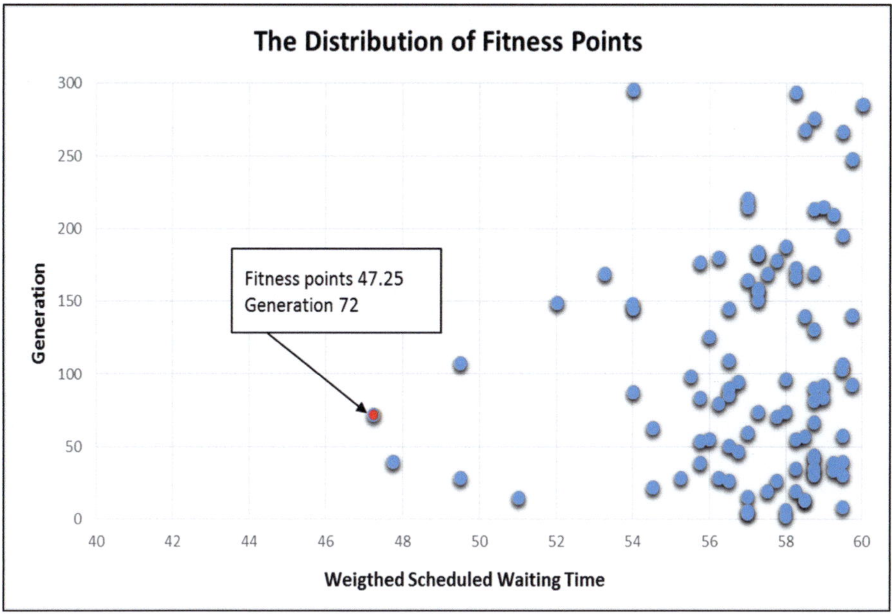

Figure 16:   **Distribution Chart of Fitness Points for Target Fitness 60**

During 18 hours of operation, low-speed trains were overtaken by high-speed trains at all middle stations, including 101, 102, 103, 104, 105, and 106, but excluding the departure station (100) and the terminal station (107). In the optimal schedule at the target fitness of 60, there were no instances that high-speed trains and general trains are overtaken by freight trains or high-speed trains are overtaken by general trains.

Table 11 shows the number of train schedules where the same overtaking station was requested out of the 100 train schedules. There were 87 train schedules where an overtaking occurred at stations 101, 102, 103, 104, 105, and 106.

Overtaking Station	Number of Train Schedules (chromosome)
101, 102, 103, 104, 105, 106	87
101, 102, 103, 104, 106	7
101, 102, 103, 105, 106	3

Table 11:   **Number of Train Schedules where the Same Overtaking Station in Target Fitness of 60**

There were seven train schedules where an overtaking occurred at Stations 101, 102, 103, 104, and 106. There were three train schedules where an overtaking occurred at Stations 101, 102, 103, 105 and 106. The generation of the three train schedules among the performance of 100 times are analyzed as zero. That means is that it cannot find a fitness of less than 60 among 300 generations. Therefore, total number of train schedules is 97.

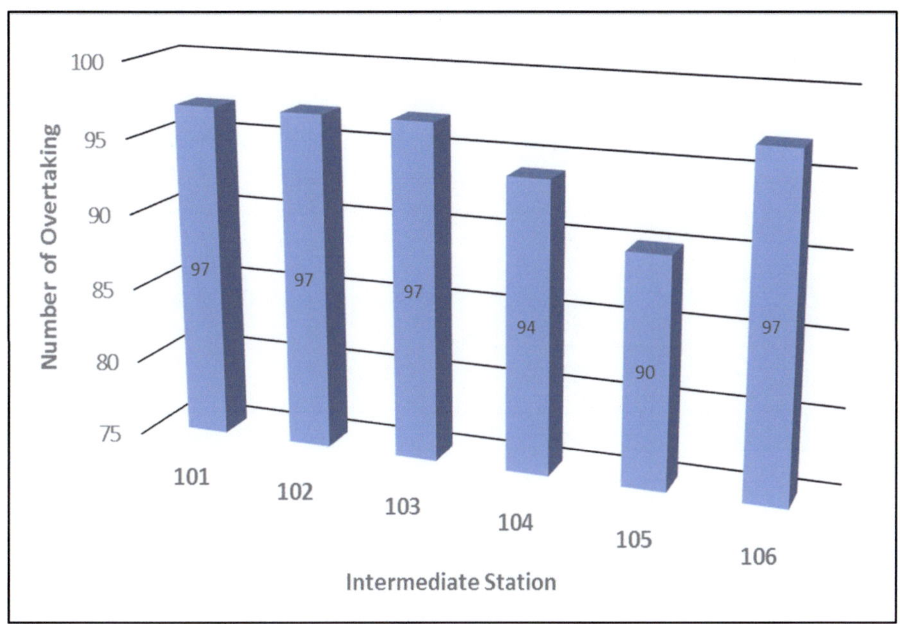

Figure 17: **Number of Overtaking on each Station for Target Fitness 60**

Figure 17 shows the number of times low-speed trains were overtaken by high-speed trains even once at each station in the 100 analyzed train schedules. Station 105 was observed to have the least overtaking at 90 times, followed by Station 104 at 94 times. The overtaking trains occurred in all 97 train schedules at Stations 101, 102, 103, and 106.

Table 12 shows the finesses of train schedules with five or less overtaking stations among six intermediate stations (101,102,103,104,105 and106) except for the departing and the terminating stations, the generations that find these fitness, and the stations where SMTs are required at the target fitness of 60.

It is analyzed that three train schedules do not require SMTs at Station 104, and seven train schedules do not require SMTs at Station 105. The lowest level of fitness for train schedules with five or less overtaking stations derived at the 74th generation at the target fitness of 60 is 57.25 weighted scheduled waiting time, and the concerned schedule does not require SMTs at Station 105. The train schedule with the best fitness of 47.25, which is derived at the 72th generation at the target fitness of 65, requires

SMTs at all the intermediate stations, therefore, the difference in fitness among these train schedules is 10 weighted scheduled waiting time.

No	Generation	Fitness	Station where Subsidiary Main Tracks (SMTs) are required for overtaking
1	140	58.5	101,102,103,105,106
2	22	59.5	
3	40	57.75	
4	44	58.75	101,102,103,104,106
5	167	58.25	
6	74	57.25	
7	90	55.75	
8	34	58.75	
9	15	59.5	
10	30	58.75	

Table 12:   **Fitness Point for 5 Overtaking Stations in Target Fitness of 60**

7.2.4   Target Fitness of 55

In Figure 18, each point shows the fitness point observed through the task of repeating the genetic algorithm 300 times, as well as the generation which found the fitness point. The fitness points were distributed between 50 and 54, and most fitness points were observed in zero generations. The target fitness of 55 has seventy-five train schedules analyzed as zero generation (detailed data is attached in the appendix). The reason is that analyzed fitness points of them are more than target fitness of 55. The fitness point of 42.75 weighted scheduled waiting time observed in the 138th generation is from the

OTS with the minimization of weighted scheduled waiting time (red point) out of 100 train schedules below the target fitness of 55.

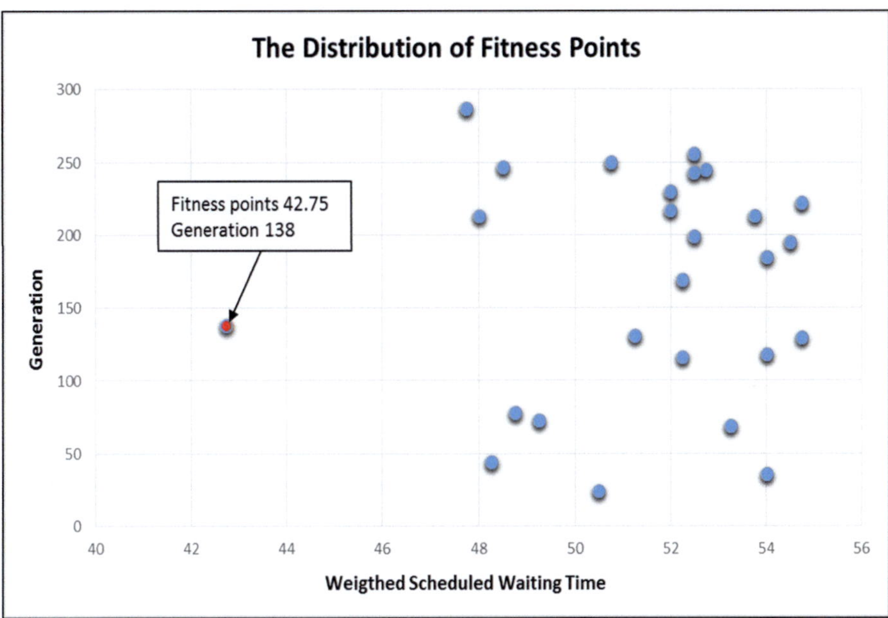

Figure 18:   **Distribution Chart of Fitness Points for Target Fitness 55**

During 18 hours of operation, low-speed trains were overtaken by high-speed trains at all middle stations, including 101, 102, 103, 104, 105, and 106, but excluding the departure station (100) and the terminal station (107). In the optimal schedule at the target fitness of 55, there were no instances that high-speed trains and general trains are overtaken by freight trains or high-speed trains are overtaken by general trains.

Overtaking station	Number of Train Schedules (chromosomes)
101, 102, 103, 104, 105, 106	23
101, 102, 103, 104, 106	3

Table 13:   **[Number of Train Schedules where the Same Overtaking Station in Target Fitness of 55]**

Table 13 shows the number of train schedules where the same overtaking station was requested out of the 100 train schedules. There were twenty-three train schedules where the overtaking occurred at Stations 101, 102, 103, 104, 105, and 106. There were two train schedules where the overtaking occurred at Stations 101, 102, 103, 104, and 106. The generation of the seventy-four train schedules among the performance of 100 times are analyzed as zero. That means is that it cannot find a fitness of less than 55 among 300 generations. Therefore, total number of train schedules is 26. The analysis is not performed on the target fitness of 50 as the number of analyzed schedules within the 300 generations at the target fitness of 55 decreases significantly.

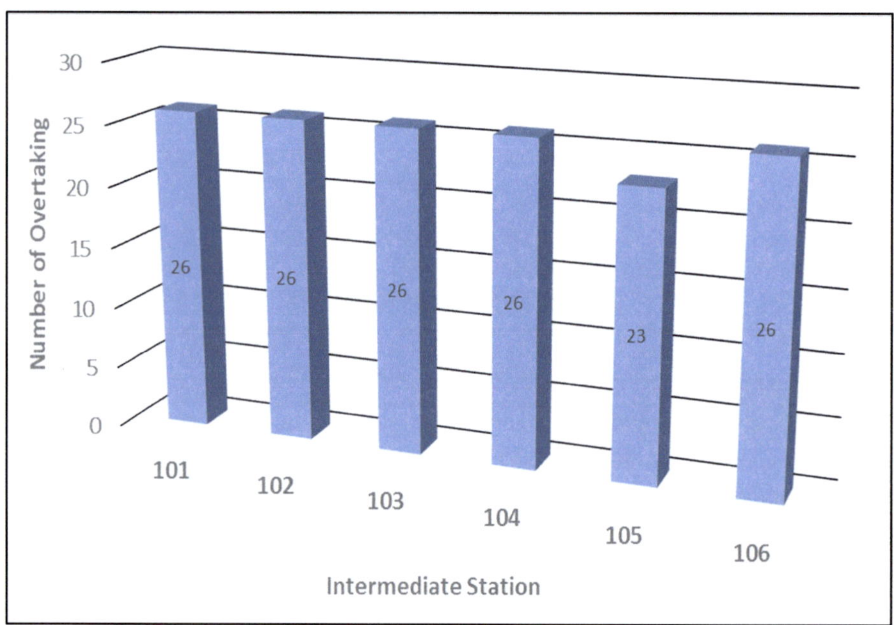

Figure 19:   **Number of Overtaking on each Station for Target Fitness 55**

Figure 19 shows the number of times low-speed trains were overtaken by high-speed trains even once at each station in the 100 analyzed train schedules. Station 105 was observed to have the least overtaking at 23 times. Overtaking trains occurred in all 26 train schedules at Stations 101, 102, 103, 104 and 106.

Table 14 shows the finesses of train schedules with five or less overtaking stations among six intermediate stations (101,102,103,104,105,106) except for the departing and the terminating stations, the generations that find these fitness, and the stations where SMTs are required at the target fitness of 55. It is analyzed that three schedules do not require SMTs at Station 105. At the target fitness of 55, the lowest level of fitness for train schedules with five or less overtaking stations is 42.75 weighted scheduled waiting time, which derived at 138th generation, and the concerned schedule does not require SMTs at Station 105.

No	Generation	Fitness	Station where Subsidiary Main Tracks (SMTs) are required for overtaking
1	169	52.25	
2	213	48	101,102,103,104,106
3	138	42.75	

Table 14:   **Fitness Point for 5 Overtaking Stations in Target Fitness of 55**

7.3   Analysis Results

Table 15 shows the best fitness for each target fitness analysis and the stations where high-speed trains overtake low-speed (general and freight) trains on train schedules to meet this best fitness. The best fitness of the target fitness 70 and 65 is 47 weighted scheduled waiting time. The best fitness of the target fitness of 60 is 47.25 weighted scheduled waiting time, and the difference between this and that of the aforementioned target fitness is 0.5 weighted scheduled waiting time.

Division	Target Fitness			
	70	65	60	55
Best fitness	47.75	47.75	47.25	42.75
Generation	4	29	72	138
Overtaking Station	101,102,103, 104,105,106	101,102,103, 104,105,106	101,102,103, 104,105,106	101,102,103, 104,106

Table 15:   **Comparison of the Best fitness for each Target Fitness**

The 0.5 weighted scheduled waiting time is considered to be no big difference considering the dwell time for general trains of one minute. When the target fitness is 55, the best fitness is 42.75 weighted scheduled waiting time, which shows 4.5 weighted scheduled waiting time difference from the best fitness of 47.25 weighted scheduled waiting time at the target fitness 60. While all six stations require overtaking facilities at the target fitness of 70, 65 or 60, only five stations require overtaking facilities and the Station 105 does not require Subsidiary Main Tracks (SMTs) at the target fitness of 55. The lower the target finesses are, the more likely generations are to find the best fitness.

The target fitness represents the overall schedule waiting time of the train schedule. A low target fitness means that the criteria for overall schedule waiting time in the train schedule is low. A target fitness with a low overall schedule waiting time has more strict conditions. The lower the target finesses are, the more likely the algorithms designed in this study are to repeat the generation. When the constraints are stringent, the number of the repetition of this task (generation) increases. In other words, there are more iterations of evaluation, selection, transition, and repair in order to derive the optimal solution that satisfies the constraints in the optimized algorithm. Also, the number of train operations and stop stations make the process of finding the optimal solution complicated, resulting in a relatively longer analysis time.

Table 16 shows the best finesses and overtaking stations of the schedules that require SMTs at five stations in the train schedules

Division	Target Fitness			
	70	65	60	55
Best fitness	62.25	53	57.25	42.75
Generation	9	2	74	138
Overtaking Station	101,102,103, 105,106	101,102,103, 105,106	101,102,103, 104,106	101,102,103, 104,106

Table 16: **The Best fitness and Under Restriction of Five Stations that Require Subsidiary Main Tracks (SMTs)**

When the target fitness is 70, the best fitness of 62.25 weighted scheduled waiting time is analyzed in the 9 generations, and this train schedule (chromosome) does not need SMTs at Station 104. At the target fitness of 65, the best fitness of 53 weighted scheduled waiting time is analyzed in the 2 generations. As in the case of the target fitness 70, this train schedule does not require SMTs at Station 104. This means that high-speed trains do not overtake low-speed (general and freight) trains in the 20.47 km section between Station 103 and Station 105.

At the target fitness of 60, the best fitness of 57.25 weighted scheduled waiting time is derived at the 74 generations, and SMTs are not necessary at Station 105 according to this train schedule. Lastly, the best fitness of 42.75 weighted scheduled waiting time is analyzed at the 138 generations when the target fitness is 55. The same as the case of target fitness of 60, for this train schedule, SMTs are not a necessity. This indicates that high-speed trains do not outstrip low-speed trains in the 18.74 km section between Station 104 and Station 106 based on this train schedule.

Division	Target Fitness							
	70		65		60		55	
Station	Number of overtaking	Ratio	Number of overtaking	Ratio	Number of overtaking	Ratio	Number of overtaking	Ratio
101	100	1	100	1	97	1	26	1
102	100	1	100	1	97	1	26	1
103	100	1	100	1	97	1	26	1
104	98	0.98	93	0.93	94	0.97	26	1
105	99	0.99	86	0.86	90	0.93	23	0.88
106	100	1	100	1	97	1	26	1

Table 17:   The Number of Overtaking and Ratio for Each Station among Analyzed 100 Train Schedules

Table 17 shows the number of overtaking and the ratio of overtaking for each station among analyzed 100 train schedules according to each target fitness. Overtaking occurs at Station 101, 102 and 103, as well as at Station 106 based on analyzed the train schedules. At Station 104 and 105, overtaking ratio are relatively low. In particular, the 105 station shows the lowest overtaking ratio.

As a result of analyzing each target fitness, the best fitness is 42.75 weighted scheduled waiting time, and the relevant train schedule requires SMTs to all stations except for Station 105. The best fitness is 47.25 weighted scheduled waiting time at the target fitness of 60, which shows 4.5 weighted scheduled waiting time difference from the best fitness 42.75 at the target fitness of 55. The schedule having the best fitness of 47.25 weighted scheduled waiting time requires SMTs at all stations and the schedule with the best fitness of 42.75 weighted scheduled waiting time does not require SMTs at Station 105. The train schedule with the best fitness of 42.75 weighted scheduled waiting time is the nearly optimal schedule because its scheduled waiting time is at the lowest level on the route applied to this study and the number of stations that requires SMTs is 5.

The train route applied in this study is currently under basic planning and the design is under way for building optimal facilities according to a train schedule. In accordance with the most probable train schedule from the planner's view, a rail line is designed without SMTs at Station 105 as with the layout of Figure 11. This study applies the meta-heuristic algorithm to realize scheduled weighting time and the train schedule by searching more objective and empirical results. It can be said that the algorithm used in this study can actually be used in the design of the railway line.

8 Summary, Conclusion and Further Research

In this thesis, a model is presented and an algorithm is proposed to minimize weighted scheduled waiting time and solve the problem choosing the locations (stations) to be overtaken preceding low-speed trains from following high-speed trains when three different types of trains are operated on the same railway line. Furthermore, the model and algorithm satisfy the constraints listed as follows.

- Train operation frequency during calculation time
- Guaranteed arrival of the headway of high-speed train and the guaranteed departure headway of the low-speed train
- Dwell time
- Number of lines and platforms in a station
- Conditions for non-occurrence of overtaking between stations
- Running time
- Determination of a station at which to install a subsidiary main track (SMT) for overtaking
- Blocking time

Existing studies mainly used the heuristic method to solve the NP-hard problem. With this theoretical background, the problem presented in this work is defined as NP-hard and the solution of the problem was determined using a heuristic method. Thus, the genetic algorithm optimization method, which is a kind of the metaheuristic method, was applied as a train schedule optimization algorithm. This study used a genetic approach to improve the reliability of the objective functions in mathematical models, and to determine the stations that require a SMT for overtaking during the strategic and tactical stages of the railway planning process.

The data input and results output were implemented using HTML5, and the algorithm was implemented using Javascript/jQuery. Although this work aimed to obtain specific values, a more universal utilization of the results will be possible if the user interface is improved. Furthermore, it is expected that results can be obtained under complex conditions with shorter calculation time if the algorithm implementation is converted into one that is server-based.

In this thesis, a railway line was selected that plays the role of interconnecting the western region of the capital area and the Honam region in the South-North direction, and contains a high-speed railway function in preparation for the future inter-Korea rail transportation. In the short term, this line will solve the problem of insufficient line capacity of the Seoul–Busan axis by supplying high-speed trains to the Chungcheong region. In the long term, it will provide main railway functions in preparation for the connection of the inter-Korean railway. A mathematical model based on a genetic algorithm was applied to a real railway use case.

The precondition for the application of the model to analyze the Optimal Train Schedule (OTS) is that the overtaking is possible at every station and a SMT can be installed at every station that has been set. If overtaking is enabled to allow other trains to pass while a train is stopped at a specific station, then the number of trains available for operation increases because the temporal space can be used more efficiently in the link between the stations.

The observed time period is set at 18 hours, and the scheduled waiting time for the overtaking is the time between the arrival of a low-speed train at the station and the departure of the train after the high-speed train following it passes. In addition, the set limits for guaranteed arrival (1.5 minutes) and guaranteed departure (1 minute) were applied.

The genetic algorithm used in this paper analyzes the random train schedules until the given conditions are met. To obtain the study results, train schedules were analyzed for 300 generations, and 100 train schedules were calculated for the same conditions. From among these, the train schedule with the shortest scheduled waiting time was observed to be the OTS.

The train schedule derived has the best fitness of 42.75 weighted scheduled waiting time and does not require Subsidiary Main Tracks (SMTs) to be installed at Station 105 as there is no need of overtaking at this station.

In the planning of railway lines, if objective data can be provided to support the fact that a certain station require no SMTs, this plan can be regarded as highly feasible as it can lead to construction cost reduction. However, it is better to have SMTs at the station for the flexibility of train operation and the capacity of the railway line. This situation has become an issue in the planning of railroad routes.

In all train schedules, it was observed that there were no instances of high-speed trains and general trains passing freight trains or high-speed trains passing general trains. The reason for this is that the weights for passing used in the genetic algorithm analysis were 10 for general trains and 1 for freight trains in this study. The weights are applied to give high-speed trains top priority and to set priorities between general trains and freight trains when implementing algorithms. Therefore, while scheduled waiting time is not caused by other trains in the operations of high-speed trains, the scheduled waiting time occurs in the operations of general trains due to high-speed trains, and in the operations of freight trains due to high-speed trains and general trains.

As the target fitness decreased, the number of generations used to find the OTS increased. This is because a low target fitness means that the scheduled waiting time will be short. Owing to this, the analysis tasks are repeated many times when using a genetic algorithm to analyze an OTS with a minimization of weighted scheduled waiting time.

The best fitness from among the several fitness were observed to be 42.75 weighted scheduled waiting time. Moreover, the overtaking was occurred at Stations 101, 102, 103, 104, and 106 in the train schedule (chromosome). Finally, this train schedule (chromosome) was observed to have the fewest overtaking stations with the best fitness.

In order to provide maximum transportation services and improve railway management with limited railway resources, accurate and highly reliable analyses of train schedules in the railway planning must be conducted. To improve the accuracy and reliability of analysis, factors such as the effect of spatial distance, calculation of scheduled speed for various trains, train operation frequency based on traffic demand, and effect of heterogeneity of train types should be considered. Considering these variables, this work focused on the analysis of the OTS. Compared to the existing empirical method, the proposed method can examine train schedules for more scenarios, apply quantitative evaluation criteria, and review concrete infrastructures.

At present, Republic of Korea is planning various railway lines. Furthermore, the possibility of connecting the inter-Korea railway is increasing as Republic of Korea-China summit discusses a railway connection through Sinuiju in North Korea to China. It is expected that not only high-speed trains but also freight trains could reach beyond Russia and China, to Europe.

If the OTS is established by applying the method proposed in this study to the railway line currently being planned, and the stations where low-speed trains can be overtaken by high-speed trains when different types of trains operate on the same railway line are determined in accordance with the established train schedule, then it will be possible to provide an objective basis for the railway planning, rather than simply relying on the subjective experience of planners. Furthermore, if the OTS is analyzed based on the method proposed in this study when a new railway line is constructed in a developing country, then it will help achieve economic and policy feasibility.

One subject for future research is to calculate the maximum train operations that can be added depending on the ratio of train types without changing the departure times and secondary line positions of the analyzed the OTS. Based on this, increase of capacity of a line can be studied. Furthermore, although this study assumed that three types of trains are operating on the same railway line, it is necessary to implement the model for railway network and an optimization algorithm that can analyze various train operation patterns such as express and slow operation patterns in a subway, middle and long-distance express trains of a wide-area railway, and general train operation patterns for short-distance and subsidiary operations.

Finally, it is necessary to compare the genetic algorithm applied to the model proposed in this study, with other metaheuristic algorithms to examine the adequacy of the genetic algorithm.

Glossary

Algorithm	A set of step-by-step instructions for solving a problem or a well-defined procedure to find a feasible solution to an instance of a problem in a finite number of steps.
Block section	A section of track in a fixed block system, which a train may only enter when it is not occupied by other vehicles.
Buffer time	An extra time that is added to the minimum line headway to avoid the transmission of small delays.
Capacity	The maximum traffic flow a piece of infrastructure (line, interlocking, terminal, yard) can handle under specified operating conditions.
Chromosome	A string of genes that represent an individual.
Collateral delay	A delay that was transmitted from another train.
Constraints	In an optimization model the constraints describe which combinations of values for the decision variables are allowed.
Crossover	A reproduction operator that creates a new chromosome by exchanging parts of two existing chromosomes.
Cyclic timetable	A timetable in which trains that belong to the same route are scheduled with fixed time intervals between their train paths.
Dwell time	The total elapsed time from the time that a train stops in a station until the time it resumes moving.
Fitness	The ability of a living organism to survive and reproduce in a specific environment. Also, a value associated with a chromosome that assigns a relative merit to that chromosome.

Fitness function	A mathematical function used for calculating the fitness of a chromosome.
Generation	One iteration of a genetic algorithm.
Genetic algorithm	A type of evolutionary computation inspired by Darwin's theory of evolution. A genetic algorithm generates a population of possible solutions encoded as chromosomes, evaluates their fitness, and creates a new population by applying genetic operators – crossover and mutation. By repeating this process over many generations, the genetic algorithm breeds an optimal solution to the problem.
Heuristic	A strategy that can be applied to complex problems; It usually – but not always – yields a correct solution. Heuristics, which are developed from years of experience, are often used to reduce complex problem solving to more simple operations based on judgment. Heuristics are often expressed as rules of thumb.
Interlocking	1) Interconnection of signaling components and systems designed so that no conflicting movements can be signaled. 2) An arrangement of points and signals interconnected in a way that each movement follows the other in a proper and safe sequence.
Main tracks	The most important track among two or more tracks on which trains run in the same direction at a station.
Mutation	A genetic operator that randomly changes the gene value in a chromosome.
Objective function	An equation to be optimized given certain constraints and with variables that need to be minimized or maximized using nonlinear programming techniques. An ob-

	jective function can be the result of an attempt to express a goal in mathematical terms for use in decision analysis, operations research or optimization studies.
Optimization	An iterative process of improving the solution to a problem with respect to a specified objective function.
Overtaking	The preceding low-speed trains stop on secondary main tracks in the station so as not to obstruct the operation of the subsequent high-speed trains. After the high-speed trains passes the station, the low-speed trains can depart.
Overtaking station	A station in order that preceding low-speed trains stop on secondary main tracks in the station so as not to obstruct the operation of the subsequent high-speed trains.
Population	A group of individuals that breed together.
Reproduction	The process of creating offspring from parents.
Relief Track	A railroad track for shunting trains in the following circumstances: there is need for the following train to overtake the preceding train; there is need to allow the following train to approach prior to the departure of the preceding train, due to high train density; or, there is need to stop freight trains at the station for a long time, due to the formation and arrangement of freight trains.
Schedule	The designation of train description, day, route, speed, arrival and departure times of a train.
Scheduled waiting time	The waiting time that is needed for a scheduled passing and overtaking and to synchronize the schedules of a cyclic interval timetable.
Selection	The process of choosing parents for reproduction based on their fitness.

Shunting	Movements other than train movements accomplished at restricted speed within designated limits for making up trains, moving vehicles between tracks and similar purposes.
Subsidiary main track	A main track auxiliary to the principal main track. The principal portion of a rail transit line for use in the arrival, departure and/or passage of trains. According to its intended purpose, the main track is divided into an up track, down track, arrival track, departure track, passenger track, freight track, through track, and relief track. Most important is the principal main track, and tracks other than principal main tracks are regarded as subsidiary main tracks.
Terminal	An assemblage of facilities provided at a terminus or at intermediate points of a line for the purpose of assembling, assorting, classifying and relaying trains.
Train path	A time-distance graph that represents the schedule of a train in a traffic diagram.
Waiting for another train	The situation of train which is waiting on the other track in order to operate the upper class train or the urgent train first.

Bibliography

Abril, M.; Barber, F.; Ingolotti, L.; Salido, M. A.; Tormos, P.; Lova, A. (2008): An assessment of railway capacity. In Transportation Research Part E: Logistics and Transportation Review 44(5), pp. 774-806. DOI: 10.1016/j.tre.2007.04.001.

Ahn, Y.; Kowada, T.; Tsukaguchi, H.; Vandebona, U. (2017): Estimation of passenger flow for planning and management of railway stations. Transportation Research, Volume 25, pp 315-330. Doi.org/10.1016/j.trpro.2017.05.408

Alberto, C.; Matteo F.; Paolo T. (2002): Modeling and solving the train timetabling problem. Operation Research, Volume 50, Issue 5, pp.751-992. Doi.org/10.1287/opre.50.5.851.362.

Arenas, D.; Chevrier, R.; Hanafi, S.; Rodriguez, J (2015): Solving the train timetabling problem, a mathematical model and a genetic algorithm solution approach. 6th International Conference on Railway Operations Modelling and Analysis (Rail Tokyo 2015), Mar 2015, Tokyo, Japan.

Beheshti, Z.; S. M. H Shamsuddin. (2013): A review of population-based meta-heuristic algorithms. Int. J.Advabce. Soft Comput. Appl., Volume 5, Issue 1, pp.1-35.

Brannlund, U.; Lindberg, P. O.; Nou, A.; Nilsson, J. E. (1998): Railway Timetabling using Lagrangian Relaxation. Transportation Science, Volume 32, Issue 4, pp. 358-369. Doi.org/10.1287/trsc.32.4.358.

Burdett, R. L.; Kozan, E (2010): A disjunctive graph model and framework for constructing new train schedules. European Journal of Operational Research, Volume 200, Issue 1, pp85-98. Doi.org/10.1016/j.ejor.2008.12.005.

Burggraeve, S.; Bull, S. H.; Vansteenwegen, P.; Lusby, R. M. (2017): Integrating robust timetabling in line plan optimization for railway systems. Transportation Research Part C: Emerging Technologies, Volume 77, pp. 134-160. doi.org/10.1016/j.trc.2017.01.015.

Bussieck, M. R. (1998): Optimal lines in public rail transport. Dissertation, Technical University of Braunschweig. Faculty of Mathemaics and Computer Science.

Bussieck, M.R.; Winter, T.; Zimmerman, U.T. (1997): Discrete optimization in public rail transport. Mathematical Programming, Volume 79, Issue 1-3, pp.415-444.

Cacchiani, V.; Huisman, D.; Kidd, M.; Kroon, L.; Toth, P.; Veelenturf, L.; Wagenaar, J. (2014): An overview of recovery models and algorithms for real-time railway rescheduling. Transportation Research Part B: Methodological, Volume 63, pp. 15-37. Doi.org/10.1016/j.trb.2014.01.009.

Cao, N. (2017): Approach to determine and evaluate the influence of homogeneity of operating programs. Dissertation. Universität Stuttgart, Stutt-gart. Institut für Eisenbahn- und Verkehrswesen.

Caprara, A.; Fischetti, M.; Toth, P. (2002): Modeling and solving the train timetabling problem. Operations research, Volume 50, Issue 5, pp.851-861.

Carey, M.; Lockwood, D. (1995): A model, algorithms and strategy for train pathing. Journal of the Operational Research Society, Volume 46, Issue 8, pp. 988-1005. Doi.org/10.1057/jors.1995.136.

Chiang, T.; Hau, H. Y.; Chiang, H. M.; Kob, S. Y.; Hsieh, C. H. (1998): Knowledge-based System for Railway Scheduling. Data and Knowledge Engineering, Volume 27, Issue 3, pp. 289-312. Doi.org/10.1016/S0169-023X(97)00040-2.

Choi, G. D.; Kim, S. H. (2015): Management of train operation: An introduction to railroad engineering. Gumi Ltd, ISBN 978-89-8225-422-2, pp 479-481.

Choi, E. M. (2018): System analysis and design using UML. Life and power press. ISBN 978-89-7050-954-993560.

D' Ariano, A.; Corman, F.; Pacciarelli, D.; Pranzo, M. (2008): Reordering and local rerouting strategies to manage train traffic in real time. Transportation Science, Volume 42, Issue 4, pp.405-419. Doi.org/10.1287/trsc.1080.0247.

Do, C. U. (2013): Principles of Traffic Engineering. Cheongmungak Press, Seoul, 2013. ISBN 978-89-7088-879-8.

Fioole, P. J.; Kroon, L.; Maróti, G.; Schrijver, A. (2006): A rolling stock circulation model for combining and splitting of passenger trains. European Journal of Op- erational Research, 174 (2), 1281–1297. Doi.org/10.1016/j.ejor.2005.03.032.

Hantsch, F.; Martin, U.; Cao, N. (2016): Effiziente Infrastrukturnutzung unter Berücksichtigung der Homogenität im Eisenbahnbetrieb. Verkehrswissenschaftliche Tage, Dresden, 2016.

Hassannayebi, E.; Zegordi, S. H.; Yaghini, M. (2016): Train timetabling for an urban rail transit line using a lagrangian relaxation approach. Applied Mathematical Modelling, Volume 40, Issue 23, pp. 9892-9913. Doi.org/10.1016/j.apm.2016.06.040.

Higgins A.; Kozan E.; Ferreira L. (1996): Optimal scheduling of trains on a single line track. Transportation Research Part B: Methodological Volume 30, Issue 2, pp. 147-161. Doi.org/10.1016/0191-2615(95)00022-4.

Holland, J. (1975): Adaptation in Natural and Artificial Systems: an introductory analysis with applications to biology, control, and artificial intelligence. University of Michigan Press, Ann Arbor.

Jovanovic, D.; Harker, P. T. (1991): Tactical scheduling of rail operations: The SCAN I system. Transportation Science, Volume 25, Issue 1, pp. 46-64. Doi.org/10.1287/trsc.25.1.46.

Kim, Y. G. (2011): Evolutionary Algorithms. Chonnam National Uiversity Publishing. ISBN 978-89-7598-897-4.

Kim, Y. G. (2017a): Metaheuristics. Chonnam National Uiversity Publishing. ISBN 978-89-6849-367-6.

Kim, H. J.; Martin, U.; Hantsch, F. (2017): A study on methods of capacity research for performance evaluation. International Journal of Advanced Railway, Volume 5, Issue 2, pp. 407-419.

Kim, H. S. (2017b): Railway Allocation Modeling Using Genetic Algorithm. Dissertation. University of Seoul, Department of Transportation Engineering.

Kroon, L. G.; Huisman, D.; Maroti, G. (2007): Railway timetabling from an operations research. Report No. EI 2007-22, Econometric Institute, Erasmus University Rotterdam.

Kwan, R. S. K.; Mistry, M. (2003): A co-evolutionary algorithm for train timetabling. The 2003 Congress on Evolutionary Computation, 2003. CEC'03. Doi:10.1109/CEC.2003.1299937.

Kuznetsov, N. A.; Minashina, I. K.; Paschchenko, F.F.; Ryabykh, N. G.; Zakharova, E. M. (2014): Design and analysis of optimization algorithms for multi agent railway control system. Procedia Computer Science, Volume 37, pp 214-219. Doi.org/10.1016/j.procs.2014.08.032.

Liang, J. (2017): Metaheuristic-based dispatching optimization integrated in multi-scale simulation model of railway operation. Dissertation. Universität Stuttgart, Stutt-gart. Institut für Eisenbahn- und Verkehrswesen.

Liu L.; Dessouky M. (2017): A decomposition based hybrid heuristic algorithm for the joint passenger and freight train scheduling problem. Computers and Operations Research, Volume 87, pp 165-182.

Liebchen C.; Möhring R. H. (2007): The modeling power of the periodic event scheduling problem: railway timetables and beyond. In Algorithmic Methods for Railway Optimization, pp. 3–40.

Liu, S. Q.; Kozan, E. (2010): Scheduling trains with priorities: a no-wait blocking parallel-machine job-shop scheduling model. Transportation Science, Volume 45, Issue 2, pp.175-198. Doi.org/10.1287/trsc.1100.0332.

Lu, G.; Zhou, X.; Mahmoudi, M.; Shi, T.; Peng, Q. (2019): Optimizing resource recharging location-routing plans: A resource-space-time network modeling framework for railway locomotive refueling applications. Computers & Industrial Engineering, Volume 127, pp.1241-1258.

Lusby, R. M.; Larsen, J.; Ehrgott, M.; Ryan, D. (2011): Railway track allocation: models and methods. OR Spectrum 33, Volume 33, Issue 4, pp.843-883.

Lusby, R. M.; Larsen, J.; Bull, S. (2018): A survey on robustness in railway planning. European Journal of Operational Research 266, pp.1-15. Doi.org/10.1016/j.ejor.2017.07.044.

Martin, U. (2015): Knotenkapazität-Bewertungsverfahren für das mikroskopische Leistungsverhalten und die Engpasserkennung im spurgeführten Verkehr (RePlan). Norderstedt: BoD (Neues verkehrswissenschaftliches Journal, 8)

Martin, U.; Li, X.; Warninghoff, C.-R. (2012): Bewertungsverfahren für Knotenelemente bei der Infrastrukturbemessung - Replan. In Eisenbahntechnische Rundschau (ETR) 11 (61), pp. 38–43.

Martin, U.; Liang, J. (2017): The Influence of dispatching on the relationship between capacity and operation quality of railway systems. Final Report of DFG Project (MA 2326/15-1). Norderstedt: BoD (Neues verkehrswissenschaftliches Journal, Ausgabe 20).

Moon, B. Y. (2008): An easy-to-learn genetic algorithm (evolutionary approach). Hanbit media. ISBN 978-89-7914-576-2.

Nirmala, G.; Ramprasad, D. (2014): A genetic algorithm based railway scheduling model. International Journal of Science and Research (IJSR), Volume 3, Issue 1, pp.11-14.

Oh, D. G. (2012): Optimal line planning for intercity multi-class railway. Dissertation, Seoul National University, Department of Construction and Environmental Engineering.

Omid, G.; Yuri, N. S. (2012): Train routing and timetabling via a genetic algorithm. IFAC Proceedings Volumes, Volume 45, Issue 6, pp.158-163. Doi.org/10.3182/20120523-3-RO-2023.00294.

Ortúzar, J, D.; Willumsen, L. G. (2001): Modelling Transport: Modal split and direct demand models. Third Edition. John Wiley & Sons, Ltd. ISBN 0-471-861103, pp 199-215.

Pachl, J. (2002): Railway Operation and Control. VTD Rail Publishing.

Pachl, J. (2014): Railway Timetable and Traffic. 2nd Edition. Eurailpress, DVV Rail Media (DVV Media Group GmbH), Hamburg.

Rezanova, N. J; Ryan, D. M. (2010): The train driver recovery problem- A set partitioning based model and solution method. Computers & Operation Research, Volume 37, Issue 5, pp.845-856. Doi.org/10.1016/j.cor.2009.03.023.

Schmidt M.; Schöbel A. (2015): Timetabling with passenger routing. OR Spectr, Volume 37, Issue 1, pp.75–97.

Schroeder, B. J.; Cunningham, C. M.; Findley, D J.; Hummer, J. E.; Foyle, R. S.

(2010): Manual of Transportation Engineering Studies 2nd Edition. Institute of Transportation Engineers (ITE), Publication No TB-012A, ISBN-13: 978-1-933458-53-1.

Serafini P.; Ukovich W. (1989): A mathematical model for periodic scheduling problems. Society for Industrial and Applied Mathematics, Volume 2, Issue 4, pp. 550–581. Doi.org/10.1137/0402049.

Shakibayifar, M.; Hassannayebi, E.; Mirzahossein, H.; Zohrabnia, S.; Shahabi, A. (2017): An integrated train scheduling and infrastructure development model in railway networks. Transactions E: Industrial Engineering. Scientia Iranica, Volume 24, Issue 6, pp 3409-3422. Doi: 10.24200/sci.2017.4397.

Tormos, P.; Lova, A.; Barber, F.; Ingolotti, L.; Abril, M.; Salido, M. A. (2008): A genetic algorithm for railway scheduling problems. Metaheuristics for Scheduling in Industrial and Manufacturing Applications, pp.256-276.

Törnquist, J. (2006): Computer-based decision support for railway traffic scheduling and dispatching: a review of models and algorithms. Dagstuhl Research Online Publication Server (DROPS)

Vaidyanathan, B.; Ahuja, R. K.; Orlin, J. B. (2008): The locomotive routing problem. Transportation Science, Volume 42, Issue 4, pp.492-507. Doi.org/10.1287/trsc.1080.0244.

Wang, J. (2018): Chapter 9: Train operation control and intelligent dispatching. In Safety Theory and Control Technology of High-Speed Train Operation, pp. 267-294. Doi.org/10.1016/B978-0-12-813304-0.00009-8.

Wegele, S.; Schnieder, E. (2004): Dispatching of train operations using genetic algorithms. Computers in Railways IX, Dresden, Germany, pp.775-784.

Wendler, E. (2007): The scheduled waiting time on railway lines. Transportation Research Part B, Volume 41, Issue 2, pp.148-158.

Zimmermann, U.; Lindner, T. (2003): Train schedule optimization in public rail transport. Mathematics Key technology for the future, Springer, pp.703-716.

List of Abbreviations

AT	Arrival time
DT	Departure time
FTS	Freight train schedule
FT	Fright train
GA	Guaranteed arrival
GD	Guaranteed departure
GT	General train
HBL	Homogeneity of blocking time
HBU	Homogeneity of buffer time
HRD	Homogeneity of running direction
HT	High-speed train
NP	Non-deterministic polynomial-time
OTS	Optimal train schedule
PT	Passenger Train
PTS	Passenger Train Schedule
SMTs	Subsidiary main tracks
TOCs	Train operating companies
TPS	Train Performance Simulation
TTP	Train timetable problem
UML	Unified modeling language

List of Variables

$a_n t_k^h$	The arrival time of kth high-speed train at a node.
$a_n t_k^g$	The arrival time of kth general train at a node.
$a_n t_k^f$	The arrival time of kth freight train at a node.
$aHw_n t_{k\to k}^{h\to g}$	Arrive headway between preceding k^{th} high-speed train and following k^{th} general train at n^{th} node.
$aHw_n t_{k\to k}^{h\to f}$	Arrive headway between preceding k^{th} high-speed train and following k^{th} freight train at n^{th} node.
$aHw_n t_{i\to j}^{g\to g}$	Arrive headway between preceding i^{th} and following j^{th} general train at n^{th} node.
$aHw_n t_{k\to k}^{g\to h}$	Arrive headway between preceding k^{th} general train and following k^{th} high-speed train at n^{th} node.
$aHw_n t_{k\to k}^{g\to f}$	Arrive headway between preceding k^{th} general train and following k^{th} freight train at n^{th} node.
$aHw_n t_{i\to j}^{f\to f}$	Arrive headway between preceding i^{th} and following j^{th} freight train at n^{th} node.
$aHw_n t_{k\to k}^{f\to h}$	Arrive headway between preceding k^{th} freight train and following k^{th} high-speed train at n^{th} node.
$aHw_n t_{k\to k}^{f\to g}$	Arrive headway between preceding k^{th} freight train and following k^{th} general train at n^{th} node.
$aHw_n t_{k\to k}^{f\to h}$	Safe arrival headway for the overtaking between preceding k^{th} freight train and following k^{th} high-speed train at n^{th} node.
$aHw_n t_{k\to k}^{g\to h}$	Safe arrival headway for the overtaking between preceding k^{th} general train and following k^{th} high-speed train at n^{th} node.
$aHw_n t_{k\to k}^{f\to g}$	Safe arrival headway for the overtaking between preceding k^{th} freight train and following k^{th} general train at n^{th} node.

$bbtt^{h}_{k,bs(k)}$	Begin of blocking time of kth high-speed train in block section K
bt	Blocking time
$d_n t^{h}_{k}$	The departure time of kth high-speed train at a node.
$d_n t^{g}_{k}$	The departure time of kth general train at a node.
$d_n t^{f}_{k}$	The departure time of kth freight train at a node.
$dHw_n t^{h \to h}_{i \to j}$	Departure headway between preceding i^{th} and following j^{th} high-speed train at a node.
$dHw_n t^{h \to g}_{k \to k}$	Departure headway between preceding k^{th} high-speed train and following k^{th} general train at n^{th} node.
$dHw_n t^{h \to f}_{k \to k}$	Departure headway between preceding k^{th} high-speed train and following k^{th} freight train at n^{th} node.
$dHw_n t^{g \to g}_{i \to j}$	Departure headway between preceding i^{th} and following j^{th} general train at n^{th} node.
$dHw_n t^{g \to h}_{k \to k}$	Departure headway between preceding k^{th} general train and following k^{th} high-speed train at n^{th} node.
$dHw_n t^{g \to f}_{k \to k}$	Departure headway between preceding k^{th} general train and following k^{th} freight train at n^{th} node.
$dHw_n t^{f \to f}_{i \to j}$	Departure headway between preceding i^{th} and following j^{th} freight train at n^{th} node.
$dHw_n t^{f \to h}_{k \to k}$	Departure headway between preceding k^{th} freight train and following k^{th} high-speed train at n^{th} node.
$dHw_n t^{f \to g}_{k \to k}$	Departure headway between preceding k^{th} freight train and following k^{th} general train at n^{th} node.
$dHw_n t^{h \to f}_{k \to k}$	Safe departure headway for overtaking between preceding k^{th} high-speed train and following k^{th} freight train at n^{th} node.

$dHw_n t_{k \to k}^{h \to g}$	Safe departure headway for overtaking between preceding k^{th} high-speed train and following k^{th} general train at n^{th} node.
$dHw_n t_{k \to k}^{g \to f}$	Safe departure headway for overtaking between preceding k^{th} general train and overtaking k^{th} freight train at n^{th} node.
$dw_n t_k^h$	Dwell time of kth high-speed train at a node.
$dw_n t_k^g$	Dwell time of kth general train at a node.
$dw_n t_k^f$	Dwell time of kth freight train at a node.
$ebtt_{k,bs(k)}^f$	End of blocking time of kth freight train in block section K
Hw	Headway.
$Hw\Delta_{n,n+1} t_{i \to j}^{h \to h}$	Headway for preceding i^{th} and following j^{th} high-speed train between n^{th} node and n^{th+1} node.
$Hw\Delta_{n,n+1} t_{k \to k}^{h \to g}$	Headway for preceding k^{th} high-speed train and following k^{th} general train between n^{th} node and n^{th+1} node.
$Hw\Delta_{n,n+1} t_{k \to k}^{h \to f}$	Headway for preceding k^{th} high-speed train and following k^{th} freight train between n^{th} node and n^{th+1} node.
$Hw\Delta_{n,n+1} t_{i \to j}^{g \to g}$	Headway for preceding i^{th} and following j^{th} general train between n^{th} node and n^{th+1} node.
$Hw\Delta_{n,n+1} t_{k \to k}^{g \to h}$	Headway for preceding k^{th} general train and following k^{th} high-speed train between n^{th} node and n^{th+1} node.
$Hw\Delta_{n,n+1} t_{k \to k}^{g \to f}$	Headway for preceding k^{th} general train and following k^{th} freight train between n^{th} node and n^{th+1} node.
$Hw\Delta_{n,n+1} t_{k \to k}^{f \to f}$	Headway for preceding i^{th} and following j^{th} freight train between n^{th} node and n^{th+1} node.
$Hw\Delta_{n,n+1} t_{k \to k}^{f \to h}$	Headway for preceding k^{th} freight train and following k^{th} high-speed train between n^{th} node and n^{th+1} node.

$\mathrm{Hw}\Delta_{n,n+1}t_{k\to k}^{f\to g}$ — Headway for preceding k^{th} freight train and following k^{th} general train between n^{th} node and n^{th+1} node.

St_n — Number of lines in a node.

Sp_n — Number of platforms in a node.

t_k^f — The kth freight train on a railway line.

t_k^h — The kth high-speed train on a railway line.

t_k^g — The kth general train on a railway line.

t_k^p — The kth proceeding train on a railway line.

t_k^s — The kth subsequent train on a railway line.

$tt_k^h\Delta_{n,n+1}$ — Running time of kth high-speed train between n^{th} node and n^{th+1} node.

$tt_k^g\Delta_{n,n+1}$ — Running time of kth general train between n^{th} node and n^{th+1} node.

$tt_k^f\Delta_{n,n+1}$ — Running time of kth freight train between n^{th} node and n^{th+1} node.

$t\Delta t_k^h$ — Conveyance time of kth high-speed train on a railway line.

$t\Delta t_k^g$ — Conveyance time of kth general train on a railway line.

$t\Delta t_k^f$ — Conveyance time of kth freight train on a railway line.

δ_n — An extra time that is added to the headway to avoid the transmission of small delays.

Appendix I: The Detailed Data of Each Target Fitness

If a simulation run, out of 100 simulation runs, did not meet the target fitness, the field for that turn was left blank.

No	Generation	Fitness	Overtaking Station	No	Generation	Fitness	Overtaking Station
1	17	64.75	101,102,103,104,105,106	51	2	68.5	101,102,103,104,105,106
2	1	69.5	101,102,103,104,105,106	52	11	60	101,102,103,104,105,106
3	7	69.5	101,102,103,104,105,106	53	12	61.25	101,102,103,104,105,106
4	3	65	101,102,103,104,105,106	54	15	67	101,102,103,104,105,106
5	9	67.25	101,102,103,104,105,106	55	3	66	101,102,103,104,105,106
6	13	69.25	101,102,103,104,105,106	56	1	66.5	101,102,103,104,105,106
7	6	64.25	101,102,103,104,105,106	57	24	69	101,102,103,104,105,106
8	27	66.25	101,102,103,104,105,106	58	1	65	101,102,103,104,105,106
9	4	68.25	101,102,103,104,105,106	59	1	68	101,102,103,104,105,106
10	1	61.25	101,102,103,104,105,106	60	16	66	101,102,103,104,105,106
11	15	63.5	101,102,103,104,105,106	61	3	68.75	101,102,103,104,105,106
12	2	67	101,102,103,104,105,106	62	4	57.75	101,102,103,104,105,106
13	25	52.5	101,102,103,104,105,106	63	11	66.25	101,102,103,104,105,106
14	25	64	101,102,103,104,105,106	64	18	69	101,102,103,104,105,106
15	16	64.75	101,102,103,104,105,106	65	10	68	101,102,103,104,105,106
16	2	68	101,102,103,104,105,106	66	15	68.75	101,102,103,104,105,106
17	7	68	101,102,103,104,105,106	67	4	69.75	101,102,103,104,105,106
18	1	68.25	101,102,103,104,105,106	68	16	64.5	101,102,103,104,105,106
19	8	58	101,102,103,104,105,106	69	11	68.25	101,102,103,104,105,106
20	5	66.75	101,102,103,104,105,106	70	4	68.75	101,102,103,104,105,106
21	2	69	101,102,103,104,105,106	71	18	63.75	101,102,103,104,105,106
22	27	66	101,102,103,104,105,106	72	10	65.75	101,102,103,104,105,106
23	9	67	101,102,103,104,105,106	73	12	66.75	101,102,103,104,105,106
24	24	69.25	101,102,103,104,105,106	74	21	64.5	101,102,103,104,105,106
25	21	66.5	101,102,103,104,105,106	75	4	67	101,102,103,104,105,106
26	15	67.75	101,102,103,104,105,106	76	13	68.75	101,102,103,104,105,106

27	14	69.5	101,102,103,104,105,106	77	1	67.75	101,102,103,104,105,106
28	8	59.75	101,102,103,104,105,106	78	3	67.75	101,102,103,104,105,106
29	17	63.25	101,102,103,104,105,106	79	6	68.5	101,102,103,104,105,106
30	99	66.75	101,102,103,104,105,106	80	9	62.25	101,102,103,105,106
31	2	65.5	101,102,103,104,105,106	81	13	67.25	101,102,103,104,105,106
32	3	68	101,102,103,104,105,106	82	14	67.75	101,102,103,104,105,106
33	2	68.5	101,102,103,104,105,106	83	8	68.5	101,102,103,104,105,106
34	20	52.25	101,102,103,104,105,106	84	7	68.75	101,102,103,104,105,106
35	11	63	101,102,103,104,105,106	85	9	68.75	101,102,103,104,105,106
36	18	67.25	101,102,103,104,105,106	86	8	63.25	101,102,103,104,105,106
37	7	62	101,102,103,104,105,106	87	1	64.5	101,102,103,104,105,106
38	8	64.75	101,102,103,104,105,106	88	10	69.25	101,102,103,104,105,106
39	12	66.25	101,102,103,104,105,106	89	13	61	101,102,103,104,105,106
40	11	65.25	101,102,103,104,105,106	90	25	67.75	101,102,103,104,105,106
41	2	68.5	101,102,103,104,105,106	91	9	69.75	101,102,103,104,105,106
42	14	69.25	101,102,103,104,105,106	92	1	68.25	101,102,103,105,106
43	14	59	101,102,103,104,105,106	93	13	68.5	101,102,103,104,105,106
44	14	65.25	101,102,103,104,105,106	94	4	69	101,102,103,104,105,106
45	7	67.75	101,102,103,104,105,106	95	1	58.25	101,102,103,104,105,106
46	4	47.75	101,102,103,104,105,106	96	10	61	101,102,103,104,105,106
47	17	62.5	101,102,103,104,105,106	97	6	67.5	101,102,103,104,105,106
48	7	69.25	101,102,103,104,105,106	98	13	67	101,102,103,104,106
49	4	54.25	101,102,103,104,105,106	99	9	68.5	101,102,103,104,105,106
50	9	67	101,102,103,104,105,106	100	5	69.5	101,102,103,104,105,106

Table 18:   **The detailed data of target fitness of 70**

No	Generation	Fitness	Overtaking Station	No	Generation	Fitness	Overtaking Station
1	1	57.75	101,102,103,104,105,106	51	57	63.75	101,102,103,104,106
2	13	58.75	101,102,103,104,105,106	52	30	56	101,102,103,104,106
3	2	63.5	101,102,103,104,105,106	53	34	63	101,102,103,104,105,106
4	5	56.5	101,102,103,104,105,106	54	39	63.25	101,102,103,104,105,106
5	33	58.75	101,102,103,104,105,106	55	19	59.25	101,102,103,104,105,106
6	9	64	101,102,103,104,105,106	56	80	62.25	101,102,103,104,105,106
7	2	53	101,102,103,105,106	57	69	62.75	101,102,103,104,105,106
8	20	62.25	101,102,103,104,105,106	58	3	59	101,102,103,104,105,106
9	6	63.25	101,102,103,104,105,106	59	18	61.75	101,102,103,104,105,106
10	69	60.25	101,102,103,104,105,106	60	49	60	101,102,103,104,106
11	10	62.5	101,102,103,104,105,106	61	28	64.75	101,102,103,104,105,106
12	9	64.25	101,102,103,104,105,106	62	1	59	101,102,103,104,105,106
13	40	61.5	101,102,103,104,105,106	63	25	62	101,102,103,104,105,106
14	53	62	101,102,103,104,106	64	17	62.75	101,102,103,104,106
15	3	62.25	101,102,103,104,105,106	65	7	57.25	101,102,103,104,106
16	60	62.5	101,102,103,104,105,106	66	17	63	101,102,103,104,105,106
17	45	64.25	101,102,103,104,105,106	67	4	64.25	101,102,103,104,105,106
18	59	64.75	101,102,103,104,105,106	68	95	54.75	101,102,103,104,106
19	52	56.5	101,102,103,104,105,106	69	76	61.25	101,102,103,104,105,106
20	78	61.25	101,102,103,104,106	70	46	64.75	101,102,103,104,105,106
21	41	64.75	101,102,103,104,105,106	71	65	53.25	101,102,103,104,105,106
22	19	61.5	101,102,103,104,105,106	72	65	58.25	101,102,103,104,105,106
23	16	63.75	101,102,103,104,105,106	73	34	62	101,102,103,104,105,106
24	9	64.5	101,102,103,104,105,106	74	2	56.5	101,102,103,104,105,106
25	25	56.25	101,102,103,104,105,106	75	16	62.5	101,102,103,104,106
26	30	64	101,102,103,104,105,106	76	8	64.25	101,102,103,104,105,106
27	69	64	101,102,103,104,105,106	77	6	62.5	101,102,103,104,105,106
28	37	58	101,102,103,105,106	78	11	62.75	101,102,103,104,105,106
29	4	62.75	101,102,103,104,105,106	79	2	64.5	101,102,103,104,106
30	65	64.75	101,102,103,104,105,106	80	16	60.25	101,102,103,104,105,106
31	24	61	101,102,103,104,105,106	81	1	60.5	101,102,103,104,105,106

32	20	62.5	101,102,103,104,105,106	82	4	64.5	101,102,103,104,105,106
33	41	63	101,102,103,104,105,106	83	36	54.25	101,102,103,104,105,106
34	29	47.75	101,102,103,104,105,106	84	52	62.5	101,102,103,104,106
35	50	60.75	101,102,103,104,105,106	85	17	64.5	101,102,103,104,105,106
36	59	61.75	101,102,103,104,105,106	86	33	59.25	101,102,103,104,105,106
37	44	62.75	101,102,103,104,105,106	87	34	59.75	101,102,103,104,105,106
38	53	63	101,102,103,104,105,106	88	29	63	101,102,103,104,106
39	87	63.75	101,102,103,104,106	89	5	62.25	101,102,103,104,105,106
40	16	60.25	101,102,103,104,105,106	90	46	64.25	101,102,103,104,105,106
41	20	61	101,102,103,105,106	91	40	64.75	101,102,103,104,105,106
42	23	64.5	101,102,103,105,106	92	34	59	101,102,103,104,105,106
43	24	61.25	101,102,103,104,105,106	93	4	62	101,102,103,104,105,106
44	3	61.5	101,102,103,104,105,106	94	1	62.75	101,102,103,104,105,106
45	13	64.5	101,102,103,104,105,106	95	26	54	101,102,103,104,105,106
46	13	51.5	101,102,103,104,105,106	96	13	57.5	101,102,103,105,106
47	15	61.25	101,102,103,104,105,106	97	7	63.5	101,102,103,104,105,106
48	65	64.5	101,102,103,104,105,106	98	20	58.5	101,102,103,105,106
49	52	63	101,102,103,104,105,106	99	88	63	101,102,103,104,105,106
50	17	63.5	101,102,103,104,106	100	79	64.5	101,102,103,104,105,106

Table 19:　　　**The detailed data of target fitness of 65**

No	Generation	Fitness	Overtaking Station	No	Generation	Fitness	Overtaking Station
1	180	56.25	101,102,103,104,105,106	51	170	58.75	101,102,103,104,105,106
2	74	58	101,102,103,104,105,106	52	0	0	-
3	97	58	101,102,103,104,105,106	53	110	56.5	101,102,103,104,105,106
4	54	55.75	101,102,103,104,105,106	54	7	58	101,102,103,104,105,106
5	44	58.75	101,102,103,104,106	55	103	59.5	101,102,103,104,105,106
6	141	59.75	101,102,103,104,105,106	56	149	52	101,102,103,104,105,106
7	20	57.5	101,102,103,104,105,106	57	27	56.5	101,102,103,104,105,106
8	14	58.5	101,102,103,104,105,106	58	215	57	101,102,103,104,105,106
9	67	58.75	101,102,103,104,105,106	59	126	56	101,102,103,104,105,106
10	88	54	101,102,103,104,105,106	60	90	56	101,102,103,104,105,106
11	178	57.75	101,102,103,104,105,106	61	9	58.75	101,102,103,104,105,106
12	210	59.25	101,102,103,104,105,106	62	84	59.5	101,102,103,104,105,106
13	182	57.25	101,102,103,104,105,106	63	90	55.75	101,102,103,104,106
14	167	58.25	101,102,103,104,106	64	60	56.5	101,102,103,104,105,106
15	107	59.5	101,102,103,104,105,106	65	145	57	101,102,103,104,105,106
16	63	54.5	101,102,103,104,105,106	66	215	54	101,102,103,104,105,106
17	55	56	101,102,103,104,105,106	67	40	59	101,102,103,104,105,106
18	31	58.75	101,102,103,104,105,106	68	22	59.5	101,102,103,105,106
19	34	59.25	101,102,103,104,105,106	69	39	54.5	101,102,103,104,105,106
20	221	57	101,102,103,104,105,106	70	71	55.75	101,102,103,104,105,106
21	74	57.25	101,102,103,104,106	71	40	57.75	101,102,103,105,106
22	39	59.25	101,102,103,104,105,106	72	80	47.75	101,102,103,104,105,106
23	86	56.5	101,102,103,104,105,106	73	92	56.25	101,102,103,104,105,106
24	82	58.75	101,102,103,104,105,106	74	29	59	101,102,103,104,105,106
25	184	57.25	101,102,103,104,105,106	75	47	55.25	101,102,103,104,105,106
26	2	58	101,102,103,104,105,106	76	6	56.75	101,102,103,104,105,106
27	35	58.75	101,102,103,104,105,106	77	177	57	101,102,103,104,105,106
28	157	57.25	101,102,103,104,105,106	78	20	55.75	101,102,103,104,105,106
29	214	58.75	101,102,103,104,105,106	79	196	58.25	101,102,103,104,105,106
30	248	59.75	101,102,103,104,105,106	80	29	59.5	101,102,103,104,105,106
31	108	49.5	101,102,103,104,105,106	81	95	49.5	101,102,103,104,105,106

32	140	58.5	101,102,103,105,106	82	55	56.75	101,102,103,104,105,106
33	93	59.75	101,102,103,104,105,106	83	131	58.25	101,102,103,104,105,106
34	72	47.25	101,102,103,104,105,106	84	34	58.75	101,102,103,104,106
35	169	53.25	101,102,103,104,105,106	85	267	58.75	101,102,103,104,105,106
36	31	59.5	101,102,103,104,105,106	86	51	59.5	101,102,103,104,105,106
37	145	56.5	101,102,103,104,105,106	87	169	56.5	101,102,103,104,105,106
38	151	57.25	101,102,103,104,105,106	88	173	57.5	101,102,103,104,105,106
39	84	59	101,102,103,104,105,106	89	29	58.25	101,102,103,104,105,106
40	148	54	101,102,103,104,105,106	90	13	56.25	101,102,103,104,105,106
41	165	57	101,102,103,104,105,106	91	268	58.5	101,102,103,104,105,106
42	27	57.75	101,102,103,104,105,106	92	296	58.5	101,102,103,104,105,106
43	4	57	101,102,103,104,105,106	93	16	54	101,102,103,104,105,106
44	57	58.5	101,102,103,104,105,106	94	58	57	101,102,103,104,105,106
45	42	58.75	101,102,103,104,105,106	95	15	59.5	101,102,103,104,106
46	99	55.5	101,102,103,104,105,106	96	276	51	101,102,103,104,105,106
47	35	58.25	101,102,103,104,105,106	97	30	58.75	101,102,103,104,106
48	159	57.25	101,102,103,104,105,106	98	0	0	-
49	188	58	101,102,103,104,105,106	99	286	57.75	101,102,103,104,105,106
50	294	58.25	101,102,103,104,105,106	100	0	0	-

Table 20:   **The detailed data of target fitness of 60**

No	Generation	Fitness	Overtaking Station	No	Generation	Fitness	Overtaking Station
1	36	54	101,102,103,104,105,106	51	0	0	-
2	185	54	101,102,103,104,105,106	52	0	0	-
3	0	0	-	53	0	0	-
4	69	53.25	101,102,103,104,105,106	54	0	0	-
5	195	54.5	101,102,103,104,105,106	55	0	0	-
6	0	0	-	56	0	0	-
7	0	0	-	57	0	0	-
8	0	0	-	58	0	0	-
9	0	0	-	59	0	0	-
10	130	54.75	101,102,103,104,105,106	60	73	49.25	101,102,103,104,105,106
11	0	0	-	61	0	0	
12	0	0	-	62	0	0	
13	0	0	-	63	0	0	
14	0	0	-	64	0	0	
15	0	0	-	65	0	0	
16	78	48.75	101,102,103,104,105,106	66	0	0	
17	243	52.5	101,102,103,104,105,106	67	0	0	-
18	245	52.75	101,102,103,104,105,106	68	0	0	-
19	24	50.5	101,102,103,104,105,106	69	0	0	-
20	0	0	-	70	250	50.75	101,102,103,104,105,106
21	0	0	-	71	222	54.75	101,102,103,104,105,106
22	0	0	-	72	0	0	-
23	0	0	-	73	44	48.25	101,102,103,104,105,106
24	0	0	-	74	217	52	101,102,103,104,105,106
25	0	0	-	75	0	0	-
26	0	0	-	76	256	52.5	101,102,103,104,105,106
27	0	0	-	77	0	0	-
28	0	0	-	78	0	0	-
29	0	0	-	79	0	0	-
30	0	0	-	80	0	0	-
31	0	0	-	81	0	0	-

32	0	0	-	82	0	0	-
33	0	0	-	83	0	0	-
34	0	0	-	84	0	0	-
35	0	0	-	85	169	52.25	101,102,103,104,106
36	0	0	-	86	0	0	-
37	0	0	-	87	0	0	-
38	0	0	-	88	213	48	101,102,103,104,106
39	287	47.75	101,102,103,104,105,106	89	247	48.5	101,102,103,104,105,106
40	0	0	-	90	0	0	-
41	0	0	-	91	0	0	-
42	0	0	-	92	138	42.75	101,102,103,104,106
43	0	0	-	93	0	0	-
44	0	0	-	94	0	0	-
45	213	53.75	101,102,103,104,105,106	95	116	52.25	101,102,103,104,105,106
46	118	54	101,102,103,104,105,106	96	0	0	-
47	0	0	-	97	0	0	-
48	199	52.5	101,102,103,104,105,106	98	0	0	-
49	0	0	-	99	131	51.25	101,102,103,104,105,106
50	0	0	-	100	230	52	101,102,103,104,105,106

Table 21:   **The detailed data of target fitness of 55**

Appendix II: Timetable for the Optimal Train Schedule and Traffic Diagram

At the target fitness of 55, the best level of fitness for train schedules with five or less overtaking stations is 42.75 weighted scheduled waiting time. The train schedule (chromosome) is the nearly Optimal Train Schedule (OTS) in this study. Detailed timetable of the nearly optimal schedule is follow as table 0-5.

The red text in the timetable means the scheduled waiting time and it means that the Subsidiary Main Track (SMT) is necessary at the station. If freight trains stops at a station and subsequent general train stops at the same station in dwell time of freight train, Subsidiary Main Tracks (SMTs) need at the station. If it is possible the freight train can depart at the station without scheduled waiting time after the general train departs, the timetable shows 0.0 as red text.

The table entries are coded using the following abbreviations:

- Station = St
- Departure = De
- Arrival = Ar
- Freight train = Fr
- General train = Ge
- High-speed train = Hs

[unit: minute]

St	100	101		102		103		104		105		106		107
	De	Ar	De	Ar	De	Ar	De	Ar	De	Ar	De	Ar	De	Ar
Fr	0.0	17.25	17.25	23.25	23.25	35.5	55.5 0.0	69.25	69.25	77.0	77.0	83.0	103.0 0.0	106.25
Ge	15.0	25.75	26.75	31.0	32.0	39.75	40.75	49.25	50.25	55.75	56.75	60.5	61.5	63.5
Ge	43.5	54.25	55.25	59.5	60.5	68.25	69.25	77.75	78.75	84.25	85.25	89.0	90.0	92.0
Fr	63.0	80.25	80.25	86.25	89.0 2.75	101.25	121.25	135.0	135.0	142.75	142.75	148.75	168.75 0.0	172.0
Hs	78.5	85.75	85.75	88.0	88.0	93.25	95.25	101.25	101.25	104.0	104.0	106.25	106.25	108.0

Next page continually

St	100	101		102		103		104		105		106		107
	De	Ar	De	Ar	De	Ar	De	Ar	De	Ar	De	Ar	De	Ar
Ge	106.5	117.25	118.25	122.5	123.5	131.25	132.25	140.75	141.75	147.25	148.25	152.0	153.0	155.0
Hs	144.5	151.75	151.75	154.0	154.0	159.25	161.25	167.25	167.25	170.0	170.0	172.25	172.25	174.0
Ge	173.0	183.75	184.75	189.0	190.0	197.75	198.75	207.25	208.25	213.75	214.75	218.5	219.5	221.5
Fr	184.5	201.75	201.75	207.75	207.75	220.0	240.0 0.0	253.75	253.75	261.5	261.5	267.5	288.75 1.25	292.0
Hs	208.0	215.25	215.25	217.5	217.5	222.75	224.75	230.75	230.75	233.5	233.5	235.75	235.75	237.5
Fr	224.75	242.0	242.0	248.0	248.0	260.25	280.25 0.0	294.0	297.75 3.75	305.5	305.5	311.5	331.5	334.75
Ge	241.25	252.0	253.0	257.25	258.25	266.0	267.0	275.5	276.5	282.0	283.0	286.75	287.75	289.75
Fr	257.75	275.0	275.0	281.0	284.50 3.50	296.75	316.75	330.50	330.50	338.25	338.25	344.25	364.25 0.0	367.50
Hs	274.0	281.25	281.25	283.5	283.5	288.75	290.75	296.75	296.75	299.5	299.5	301.75	301.75	303.5
Fr	284.75	302.0	302.0	308.0	308.0	320.25	340.25 0.0	354.0	358.0 4.0	365.75	365.75	371.75	391.75	395.0
Ge	303.25	314.0	315..0	319.25	320.25	328.0	329.0	337.5	338.5	344.0	345.0	348.75	349.75	351.75
Hs	334.25	341.5	341.5	343.75	343.75	349.0	351.0	357.0	357.0	359.75	359.75	362.0	362.0	363.75
Fr	356.75	374.0	374.0	380.0	380.0	392.25	412.75 0.50	426.5	426.5	434.25	434.25	440.25	460.25	463.5
Ge	369.0	379.75	380.75	385.0	386.0	393.75	394.75	403.25	404.25	409.75	410.75	414.5	415.5	417.5
Fr	379.75	397.0	397.0	403.0	405.5 2.5	417.75	437.75	451.5	451.5	459.25	459.25	465.25	485.25 0.0	488.5
Hs	395.0	402.25	402.25	404.5	404.5	409.75	411.75	417.75	417.75	420.5	420.5	422.75	422.75	424.5
Fr	412.75	430.0	430.0	436.0	436.0	448.25	473.5 5.25	487.25	487.25	495.0	495.0	501.0	521.0	524.25
Ge	426.25	437.0	438.0	442.25	443.25	451.0	452.0	460.5	461.5	467.0	468.0	471.75	472.75	474.75
Hs	455.75	463.0	463.0	465.25	465.25	470.5	472.5	478.5	478.5	481.25	481.25	483.5	483.5	485.25
Ge	490.0	500.75	501.75	506.0	507.0	514.75	515.75	524.25	525.25	530.75	531.75	535.5	536.5	538.5
Fr	507.5	524.75	524.75	530.75	533.25 2.5	545.5	565.5	579.25	579.25	587.0	587.0	593.0	613.0 0.0	616.25

Next page continually

St	100	101		102		103		104		105		106		107
	De	Ar	De	Ar	De	Ar	De	Ar	De	Ar	De	Ar	De	Ar
Hs	522.75	530.0	530.0	532.25	532.25	537.5	539.5	545.5	545.5	548.25	548.25	550.5	550.5	552.25
Fr	532.75	550.0	550.0	556.0	556.0	568.25	588.25 / 0.0	602.0	602.0	609.75	609.75	615.75	635.75 / 0.0	639.0
Ge	557.75	568.5	569.5	573.75	574.75	582.5	583.5	592.0	593.0	598.5	599.5	603.25	604.25	606.25
Fr	567.75	585.0	585.0	591.0	591.0	603.25	623.25 / 0.0	637.0	637.0	644.75	644.75	650.75	670.75 / 0.0	674.0
Hs	590.0	597.25	597.25	599.5	599.5	604.75	606.75	612.75	612.75	615.5	615.5	617.75	617.75	619.5
Fr	606.0	623.25	623.25	629.25	629.25	641.5	667.25 / 5.75	681.0	681.0	688.75	688.75	694.75	714.75	718.0
Ge	619.25	630.0	631.0	635.25	636.25	644.0	645.0	653.5	654.5	660.0	661.0	664.75	665.75	667.75
Hs	649.5	656.75	656.75	659.0	659.0	664.25	666.25	672.25	672.25	675.0	675.0	677.25	677.25	679.0
Fr	672.25	689.5	689.5	695.5	695.5	707.75	727.75 / 0.0	741.5	741.5	749.25	749.25	755.25	775.25	778.5
Ge	687.5	698.25	699.25	703.5	704.5	712.25	713.25	721.75	722.75	728.25	729.25	733.0	734.0	736.0
Hs	709.75	717.0	7170	719.25	719.25	724.5	726.5	732.5	732.5	735.25	735.25	737.5	737.5	739.25
Fr	734.0	751.25	751.25	757.25	757.25	769.5	790.75 / 1.25	804.5	804.5	812.25	812.25	818.25	838.25	841.5
Ge	747.75	758.5	759.5	763.75	764.75	772.5	773.5	782.0	783.0	788.5	789.5	793.25	794.25	796.25
Fr	757.75	775.0	775.0	781.0	783.5 / 2.50	795.75	815.75	829.5	829.5	837.25	837.25	843.25	863.25 / 0.0	866.5
Hs	773.0	780.25	780.25	782.5	782.5	787.75	789.75	795.75	795.75	798.5	798.5	800.75	800.75	802.5
Fr	787.5	804.75	804.75	810.75	810.75	823.0	843.0 / 0.0	856.75	856.75	864.5	864.5	870.5	890.5 / 0.0	893.75
Ge	805.75	816.5	817.5	821.75	822.75	830.5	831.5	840.0	841.0	846.5	847.5	851.25	852.25	854.25
Fr	822.25	839.5	839.5	845.5	845.5	857.75	877.75 / 0.0	891.5	891.5	899.25	899.25	905.25	925.25 / 0.0	928.5
Fr	833.25	850.5	854.0 / 3.5	860.0	860.0	872.25	892.25	906.0	906.0	913.75	913.75	919.75	939.75 / 0.0	943.0
Hs	845.75	853.0	853.0	855.25	855.25	860.5	862.5	868.5	868.5	871.25	871.25	873.5	873.5	875.25
Ge	877.25	888.0	889.0	893.25	894.25	902.0	903.0	911.5	912.5	918.0	919.0	922.75	923.75	925.75

Next page continually

St	100	101		102		103		104		105		106		107
	De	Ar	De	Ar	De	Ar	De	Ar	De	Ar	De	Ar	De	Ar
Hs	908.0	915.25	915.25	917.5	917.5	922.75	924.75	930.75	930.75	933.5	933.5	935.75	935.75	937.5
Fr	918.75	936.0	936.0	942.0	942.0	954.25	974.25 0.0	988.0	991.75 3.75	999.5	999.5	1005.5	1025.5	1028.75
Ge	939.75	950.5	951.5	955.75	956.75	964.5	965.5	974.0	975.0	980.5	981.5	985.25	986.25	988.25
Hs	968.0	975.25	975.25	977.5	977.5	982.75	984.75	990.75	990.75	993.5	993.5	995.75	995.75	997.5
Ge	999.5	1010.25	1011.25	1015.5	1016.5	1024.25	1025.25	1033.75	1034.75	1040.25	1041.25	1045.0	1046.0	1048.0
Ge	1032.0	1042.75	1043.75	1048.0	1049.0	1056.75	1057.75	1066.25	1067.25	1072.75	1073.75	1077.5	1078.5	1080.5
Fr	1048.0	1065.25	1065.25	1071.25	1071.25	1083.5	1103.5 0.0	1117.25	1117.25	1125.0	1125.0	1131.0	1151.0	1154.25
Ge	1060.75	1071.5	1072.5	1076.75	1077.75	1085.5	1086.5	1095.0	1096.0	1101.5	1102.5	1106.25	1107.25	1109.25
Fr	1072.25	1089.5	1089.5	1095.5	1095.5	1107.75	1127.75	1141.5	1141.5	1149.25	1149.25	1155.25	1175.25	1178.5

Table 22:   The detailed timetable of the nearly Optimal Train Schedule (OTS)

End

- **Traffic Diagram (one way)**

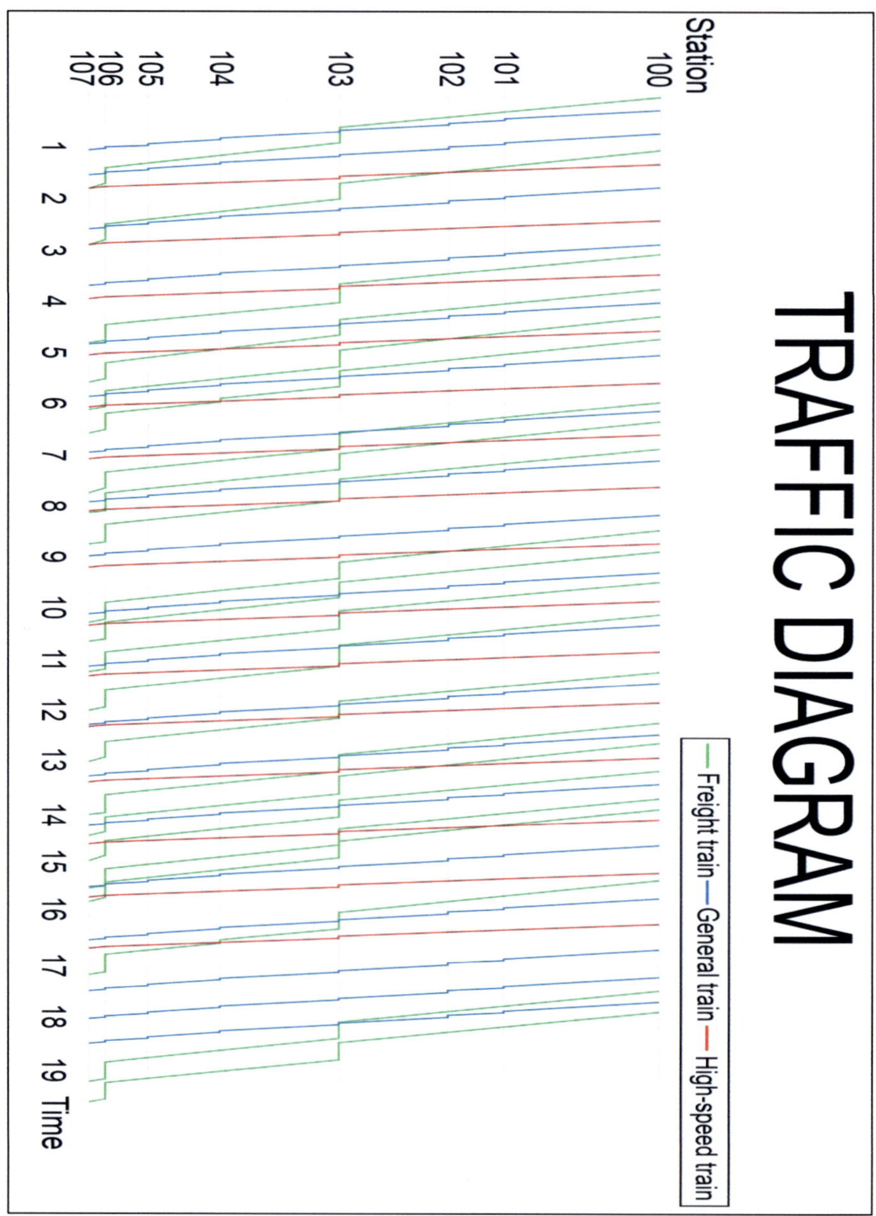

Figure 20: **Traffic Diagram of the Optimal Train Schedule (OTS) in this Study**